Well-Being Design and Frameworks for Interior Space

Valeria Minucciani
Politecnico di Torino, Italy

Nilüfer Saglar Onay
Politecnico di Torino, Italy

IGI Global
PUBLISHER of TIMELY KNOWLEDGE

A volume in the Advances in Media, Entertainment, and the Arts (AMEA) Book Series

Published in the United States of America by
IGI Global
Information Science Reference (an imprint of IGI Global)
701 E. Chocolate Avenue
Hershey PA, USA 17033
Tel: 717-533-8845
Fax: 717-533-8661
E-mail: cust@igi-global.com
Web site: http://www.igi-global.com

Library of Congress Cataloging-in-Publication Data

Names: Minucciani, Valeria, author. | Sağlar Onay, Nilüfer, author.
Title: Well-being design and frameworks for interior space / by Valeria Minucciani and Nilüfer Saglar Onay.
Description: Hershey, PA : Information Science Reference, 2020. | Includes bibliographical references and index. | Summary: "This book explores different dimensions of well-being and discusses the importance of each dimension through the examination of past and present living environments trying to figure out appropriate ways of thinking, living, and building that can lead to healthier environments and happier people"-- Provided by publisher.
Identifiers: LCCN 2020005722 (print) | LCCN 2020005723 (ebook) | ISBN 9781799842316 (h/c) | ISBN 9781799854562 (s/c) | ISBN 9781799842323 (eISBN)
Subjects: LCSH: Interior architecture--Health aspects. | Architecture--Human factors. | Well-being.
Classification: LCC NA2850 .M55 2020 (print) | LCC NA2850 (ebook) | DDC 729--dc23
LC record available at https://lccn.loc.gov/2020005722
LC ebook record available at https://lccn.loc.gov/2020005723

This book is published in the IGI Global book series Advances in Media, Entertainment, and the Arts (AMEA) (ISSN: 2475-6814; eISSN: 2475-6830)

British Cataloguing in Publication Data
A Cataloguing in Publication record for this book is available from the British Library.

Advances in Media, Entertainment, and the Arts (AMEA) Book Series

ISSN:2475-6814
EISSN:2475-6830

Editor-in-Chief: Giuseppe Amoruso, Politecnico di Milano, Italy

MISSION

Throughout time, technical and artistic cultures have integrated creative expression and innovation into industrial and craft processes. Art, entertainment and the media have provided means for societal self-expression and for economic and technical growth through creative processes.

The **Advances in Media, Entertainment, and the Arts (AMEA)** book series aims to explore current academic research in the field of artistic and design methodologies, applied arts, music, film, television, and news industries, as well as popular culture. Encompassing titles which focus on the latest research surrounding different design areas, services and strategies for communication and social innovation, cultural heritage, digital and print media, journalism, data visualization, gaming, design representation, television and film, as well as both the fine applied and performing arts, the AMEA book series is ideally suited for researchers, students, cultural theorists, and media professionals.

COVERAGE

- Visual Computing
- Communication Design
- Digital Heritage
- Cross-Media Studies
- Environmental Design
- Design Tools
- Music & Performing Arts
- Data Visualization
- Applied Arts
- Print Media

IGI Global is currently accepting manuscripts for publication within this series. To submit a proposal for a volume in this series, please contact our Acquisition Editors at Acquisitions@igi-global.com or visit: http://www.igi-global.com/publish/.

The Advances in Media, Entertainment, and the Arts (AMEA) Book Series (ISSN 2475-6814) is published by IGI Global, 701 E. Chocolate Avenue, Hershey, PA 17033-1240, USA, www.igi-global.com. This series is composed of titles available for purchase individually; each title is edited to be contextually exclusive from any other title within the series. For pricing and ordering information please visit http://www.igi-global.com/book-series/advances-media-entertainment-arts/102257. Postmaster: Send all address changes to above address.

Titles in this Series

For a list of additional titles in this series, please visit:
http://www.igi-global.com/book-series/advances-media-entertainment-arts/102257

Navigating Fake News, Alternative Facts, and Misinformation in a Post-Truth World
Kimiz Dalkir (McGill University, Canada) and Rebecca Katz (McGill University, Canada)
Information Science Reference • © 2020 • 375pp • H/C (ISBN: 9781799825432) • US $195.00

Cultural, Theoretical, and Innovative Approaches to Contemporary Interior Design
Luciano Crespi (Politecnico di Milano, Italy)
Information Science Reference • © 2020 • 459pp • H/C (ISBN: 9781799828235) • US $195.00

Handbook of Research on Combating Threats to Media Freedom and Journalist Safety
Sadia Jamil (Khalifa University, UAE)
Information Science Reference • © 2020 • 408pp • H/C (ISBN: 9781799812982) • US $265.00

Deconstructing Images of the Global South Through Media Representations and Communication
Floribert Patrick C. Endong (University of Calabar, Nigeria)
Information Science Reference • © 2020 • 469pp • H/C (ISBN: 9781522598213) • US $195.00

Handbook of Research on the Global Impacts and Roles of Immersive Media
Jacquelyn Ford Morie (All These Worlds, LLC, USA) and Kate McCallum (Bridge Arts Media, USA & Vortex Immersion Media, USA)
Information Science Reference • © 2020 • 539pp • H/C (ISBN: 9781799824336) • US $265.00

Handbook of Research on Multidisciplinary Approaches to Literacy in the Digital Age
Nurdan Oncel Taskiran (Istanbul Medipol University, Turkey)
Information Science Reference • © 2020 • 405pp • H/C (ISBN: 9781799815341) • US $265.00

For an entire list of titles in this series, please visit:
http://www.igi-global.com/book-series/advances-media-entertainment-arts/102257

IGI Global
PUBLISHER of TIMELY KNOWLEDGE

701 East Chocolate Avenue, Hershey, PA 17033, USA
Tel: 717-533-8845 x100 • Fax: 717-533-8661
E-Mail: cust@igi-global.com • www.igi-global.com

Table of Contents

Preface

THE IMPORTANCE OF THE TOPIC IN TODAY'S WORLD

The most important concrete influence on our well-being is the environment. The natural disasters that we have started to go through during the last decades clearly show that we are approaching to a point that it is probably not possible to turn back. The diversity of life on Earth is dramatically affected by human alterations of ecosystems (Baillie, Hilton-Taylor, Stuart, 2004). Compelling evidence now shows that the reverse is also true: biodiversity in the broad sense affects the properties of ecosystems and, therefore, the benefits that humans obtain from them.

Today our living environments are no more extensions of the natural environment. The solutions created by human societies, new technologies, the way they work are excessively impactful on the environment. On the other hand human societies have been built on biodiversity. Many activities indispensable for human subsistence lead to biodiversity loss, and this trend is likely to continue in the future. In addition, biodiversity has always been an integral part of the human experience. What has been less recognized is that biodiversity also influences human well-being, including the access to water and basic materials for a satisfactory life, and security in the face of environmental change, through its effects on the ecosystem processes that lie at the core of the Earth's most vital life support systems (Díaz, Fargione, Chapin, Tilman, 2006).

Although well-being in itself is subjective and it differs from person to person, at the point that we have arrived it is difficult to think of individual well-being as a process independent from the rest of the world and independent from other people and other societies. During the last centuries, in western societies the role of the individual and its independence gained more importance while in ancient times the individual's happiness was regarded as a phenomenon more related to the society. Oliver (2020), discusses that the age of the individual must end and our world depends on it. The more we see ourselves as discrete entities, the more likely we are to feel isolated and lonely and to show "selfish" behaviours. As a consequence, rates of anxiety and depression are rising across the world, while the climate and biodiversity crises deepen ever further. And the new science of social networks shows how we are linked together so closely that ideas, behaviours and preferences flow between us in a way that makes it unclear where one mind ends and another begins (Oliver, 2020). Recently also the Coronavirus has shown us the importance of the individual as a part of the society. We have seen that we can come over serious problems only if we act all together as humankind. From this perspective the health and well-being of all individuals and societies of the world depends on other individuals and societies. And there should be no discrimination between societies if we all want to be safe. So, the role and responsibility of the individual as a part of the society and the ecosystem needs to be re-evaluated as we have seen that the path that has been followed for hundreds of years has not brought positive outcomes.

The well-being of the vast majority of human societies is based more or less directly on the sustained delivery of fundamental ecosystem services, such as the production of food, fuel, and shelter, the regulation of the quality and quantity of water supply, the control of natural hazards, etc. (Díaz, Fargione, Chapin, Tilman, 2006). Recently, the bush fires in Australia, floods and cyclones in different parts of the world have shown us what can happen when we are unable to control natural disasters and its effects on nature and life species. Boyden (2016) underlines the survival of the civilization is dependent on the passage to bio-sensitive society which is in tune with, sensitive to, and respectful of the processes of life that underpin our existence. A bio-sensitive society will promote health and well-being in all sections of the human population and in the ecosystems of the biosphere. Boyden calls this cultural transformation a "cultural renaissance"or perhaps "biorenaissance" because many hunter–gatherer and early farming cultures in the past were characterized by a profound respect for the living world and an understanding that humans are part of nature and completely dependent on other people and other life forms for their well-being and survival.

Planners, architects and designers can have a great impact on our living environments and well-being. Well-being is a natural outcome of natural living not very far away from the understanding of vernacular architecture. On the other hand it is important to realize that a real and comprehensive understanding of well-being can only be achieved through the continuity of the concept to all environmental scales starting from the biosphere leading towards interiors and human bodies and also through the inclusion of all individuals and societies with all their diversities. Architecture can do a lot for well-being only if the sources of our biosphere supports our living.

THE STRUCTURE OF THE BOOK

The book is organized into 11 chapters. The first two chapters aim to discuss the concept of well-being from different perspectives including various traditions, different times in history and different contexts. Then the discussion goes in the direction of practice and policy in order to understand the legal frame about the concept. Afterwards existing design approaches that have specific value for well-being studies are discussed and their specific methodologies are analysed. Starting from Chapter 5, the authors introduce the well-being framework for interiors and apply this framework to the research and survey studies that follow this chapter.

Chapter 1, "The Well-Being Concept: Well-Being and Space," is an introduction to the concept of well-being. It examines different conceptions of well-being including the hedonic and eudaimonic traditions. After discussing the relation between design in general and well-being, the study focuses on the spatial dimensions of the concept and the role of interior design in promoting well-being and underlines the importance of the concepts of flexibility, sense of place and contact with nature.

Chapter 2, "A Short History of Well-Being in Interiors," aims to discuss the foundations of the concept of living well evaluating the understanding of "space for human well-being" in different cultural contexts at different times. Therefore, it can be seen as a very significant contribution to the modern understanding of well-being as recent well-being research has not evaluated the relation between space, human and well-being in history.

Chapter 3, "Well-Being in Practice and Policy," explores the relationship between building certification systems and people's well-being. Having therefore developed certification systems measuring and evaluating the performance of buildings in relation to their impact on the environment and its resources, we are now faced with a shift of attention on a smaller scale, linked to the performance that buildings offer not so much with respect to the environment as to the people who live in them.

Chapter 4, "Different Design Approaches for Well-Being," aims to evaluate a series of design approaches that are particularly attentive to a holistic conception of human well-being and that can be adopted with particular reference to interior design. Here it is not intended to identify a general hierarchy of such approaches, as their appropriateness depends on the context, on the user and on the interpretation of the designer himself. Therefore, it is important to know the potentials of different approaches in solving design problems with different priorities.

Chapter 5, "A Design Framework for Well-Being in Interiors," is particularly important and the chapters that follow this chapter are linked to the framework discussed in this part of the book. The environmental design framework structured here aims to contribute to sustainability with well-being criteria focusing on human factor. The framework is based on the assumption that designers need to handle the issue of well-being with a methodological approach, as it is very difficult to achieve well-being with an intuitive attitude. There are several requirements that need to be fulfilled in order to create the environment that can promote user's well-being. The framework proposal, consisting of contextual, functional, psychological, social, ergonomic, aesthetic and sensory requirements as basic design criteria, aims to support both theoretical and practical activities regarding well-being in all living environments, as a crucial component of sustainability. At the end of the chapter the hierarchy between well-being requirements is discussed through a case study run in two different cultural contexts, Italy and Turkey.

Chapter 6, "Well-Being and the Effect of Age," starts from the hypothesis that understanding of well-being evolves over time (that's during the history) and with time (that's in the lifecycle of people). Evaluating the coverage and findings of previous research about well-being and age in our living environments, it clearly emerges that the role of architecture and interior design is underestimated. The studies that focus on the relation between age and well-being lack the environmental and spatial dimension. On the other hand the studies that consider well-being related to space and age focus on only specific groups like the elderly or teenagers with problems. In this sense there is serious need for research that focus on well-being requirements related to space, age and culture. So, this chapter discusses the methodology and results of a study carried by the authors: several interviews performed in different contexts and with differently aged people have shown some differences between young people, adults and elderly.

Chapter 7, "Well-Being and Home Environments," aims to discuss the effects of each well-being requirement in home environments using the Design Framework for Well-being in Interiors discussed in Chapter 5. In this sense each requirement is evaluated in terms of its specific value for home and the discussion is enriched by a survey study that includes questions about contextual, functional, psychological, social, ergonomic, aesthetic and sensory requirements addressed in the framework. The discussion about home environments is particularly important, as the home territory is different from all other territories because of its great potential to be a place for self-expression and for freedom of action. Home also represents refugee from the outside world, a place where one can control the level of social interaction and regulate the level of privacy and interdependence. Therefore, the home territory needs to be seen as one of the most powerful promoters of well-being and designers need to work with other disciplines involved in the general research of well-being.

Chapter 8, "Well-Being and School Life," discusses the concept of well-being in school environments through a survey study carried out in two different school complexes (students aged between 11 and 14 years) in the Turin metropolitan area, where different socio-economic and cultural conditions are represented. The questionnaire submitted to the students is articulated according to the different requirements of well-being: functional, social, psychological, ergonomic, aesthetic, sensorial and connection with the context. The "school climate" includes both social and physical aspects of the school life, and can affect positively (or not) behaviour, achievements and cultural development. The attention of the students in expressing their preferences and hierarchies in the answers shows how, even in a very particular age such as pre-adolescence, the interior environment in which they spend most of their time is greatly important, and how they also have precise ideas on how much it makes their lives more comfortable and pleasant.

Chapter 9, "Cultural Components of Well-Being and Vernacular Examples," leads the discussion deeper on interiors, starting from a short survey on international declarations and charters, and discussing the strong connectedness between vernacular design principles and human well-being as an outcome. Regarding ways of living of the public, vernacular architecture holds the most important clues about interior aspects conveying living styles and attitude towards natural environment and resources. In this sense this chapter focuses on vernacular architecture as a true expression of everyday life, of great part of the public, more diffused. In particular vernacular architecture communicates the coeval concept of well-being, as it is not detached from daily life, environment and cultural values. In order to attract attention to the cultural components of well-being and their meaningful reflections on the creation of interior space, interiors of vernacular structures from selected cultural contexts are examined in terms of well-being requirements discussed in Chapter 5 and the solutions are evaluated in terms of their relation to the particular context and society.

Chapter 10, "Cultural Heritage and Reuse as a Contributor to Well-Being," aims to discuss the dialectic relationship between interior environments of heritage buildings and users, and the importance of preserving interior elements as communicators of cultural significance. In the perspective of cultural conservation, existing buildings represent the continuity of cultural values from the past to the future and at this point interior architecture is crucial as it provides an authentic interaction between users and spaces conveying all these values. Therefore the conservation of interior envelope and interior elements rises as a very important issue to be discussed, affecting the occupant's well-being in a very subjective and sometimes unconscious way (considering the cultural belonging).

Chapter 11, "Conclusion: New Perspectives for Well-Being," discusses the need for interdisciplinary research and approaches for well-being. Therefore, it is very important to investigate the intersections of well-being with architecture, medicine, social sciences as well as neuroscience. Recent well-being research shows that we have the amazing ability to reshape our brains, to change our affective mindsets, and to improve our well-being. On the other hand it is necessary to understand cultural differences and its effects on well-being and to grasp the issue in accordance with the specific context.

In general, this book tries to understand how design and architecture can contribute to the way we build up well-being. Since interior space is one of the most important determinants of our everyday experiences, its role in well-being as a conscious construct needs to be the most important concern of spatial design.

CONTRIBUTION TO THE SUBJECT MATTER

The book proposes a framework including different dimensions of well-being and discusses the importance of each dimension through the examination of past and present living environments trying to figure out the appropriate ways of thinking, living and building that can lead to healthier environments and happier people.

Regarding the environment we live in, interiors have specific importance related to the fact that interiors surround us and affect the way we behave and we feel. It is true that the natural environment can have a very important effect on our well-being but they are already there, we only need to protect them. On the other side interiors are designed to answer to people's different needs and they could be designed in countless other ways. Abercrombie (1990) discusses that when we enter a building, we cease being merely its observer; we become its content. This is why interiors have a great potential on our well-being. Previous research has shown that well-being has an environmental dimension and people's well-being can be directly and indirectly affected by the environment. But it is also important to accept that the nature of well-being has many other dimensions like the genetically determined happiness set point and other life circumstances (e.g. factors such as income, marital status, or religiosity) that cannot be controlled by the environment. On the other hand positive cognitive, behavioral, and goal-based activities can be enriched by the environment we live in and according to Lyubomirsky, Sheldon, and Schkade (2005) these activities have the potential to account for a significant portion (up to 40%) of individual differences in well-being.

The contribution of this book to interior architecture includes all basic components of the discipline's body of knowledge, which are research, practice and education. Regarding research, the most important contribution of the study is the proposal of a framework that can help to understand the nature of well-being in interiors with different purposes and also in different contexts. The framework is based on requirements, which are essential for well-being in interior space. In different chapters of the book the framework is already applied to groups of people from different age groups in different contexts and the results have been discussed in terms of well-being requirements. In the chapters based on survey studies the focus of the framework is on the human factor. On the other hand, the framework has also been applied to existing living environments to evaluate their capacity to promote well-being. In this case the focus is directly on practical examples with different qualities. These studies help to understand the influence of interior space in creating well-being for its inhabitants. Regarding practice, the framework can be a road map for designers in order to develop a comprehensive design approach for future interventions. In Chapter 4 the existing design approaches that have significant value in practice have also been discussed and their specific methodologies have been analysed. Lastly, the book offers a comprehensive understanding of well-being for education programs of interior architecture as well as architecture. For design and architecture students, it is necessary to understand the factors that have impact on people's well-being. These factors discussed all through the book include the history of the concept of living well, the evolution of well-being with age, the requirements that affect well-being, the potentials of certain design approaches for well-being, the existing environments (such as vernacular structures, heritage buildings) with specific advantages for well-being, changes in well-being requirements, interior environments with different functions (such as schools and home environments) and the intersections of interior design with other disciplines involved in the general research of well-being.

REFERENCES

Abercrombie, S. (1990). *A philosophy of interior design*. Avalon Publishing.

Baillie J.E.M., Hilton-Taylor, C., & Stuart, S.N. (2004). *IUCN Red list of threatened species: A global species assessment*. Gland, Switzerland: IUCN.

Boyden, S. (2016). The biohistorical paradigm: The early days of human ecology at the Australian National University. *Human Ecology Review*, *22*(2), 25–46. doi:10.22459/HER.22.02.2016.02

Díaz, S., Fargione, J., Chapin, F. S., & Tilman, D. (2006). Biodiversity loss threatens human well-being. *PLoS Biology*, *4*(8), e277. doi:10.1371/journal.pbio.0040277 PMID:16895442

Lyubomirsky, S., Sheldon, K. M., & Schkade, D. (2005). Pursuing happiness: The architecture of sustainable change. *Review of General Psychology*, *9*(2), 111–131. doi:10.1037/1089-2680.9.2.111

Oliver, T. (2020, January 20). *The age of the individual must end – our world depends on it*. Retrieved from www.theguardian.com/books

Saglar Onay, N., & Minucciani, V. (2018). Wellbeing Framework as a Contributor to Sustainability. In *WIT Transactions on Ecology and the Environment, Volume 2* (Vol. 17, pp. 699–710). WIT Press.

Acknowledgment

The Authors thank the Istituto Comprensivo Centro Storico of Moncalieri (Turin, Italy) that has consented to and cooperated in carrying out the survey on well-being in school environments, within the framework of a Memorandum of Understanding with the Polytechnic of Turin. They would thank in particular the Head, Prof. Valeria Fantino, and Prof. Maria Maddalena Margaria.

Chapter 1
The Well-Being Concept:
Human and Space

ABSTRACT

Well-being is related not only to physical health, but also to many other factors about our lives and living environments. It is a very ephemeral condition in relation to the complex system of interconnected components changing from culture to culture and person to person together with time and space. In this sense, design for well-being is an issue that requires intense research on subjective aspects. The chapter begins by discussing different conceptions of well-being and goes on by discussing what design can do for well-being. After discussing the spatial dimensions of well-being, the study focuses on the role of interior design in promoting well-being and underlines the importance of the concepts of flexibility, sense of place, and contact with nature.

DIFFERENT CONCEPTIONS OF WELL-BEING

Well-being is an issue studied by different disciplines like psychology, sociology, philosophy, medicine, etc. The research on well-being is mainly based on surveys asking people how satisfied they are with their lives. In this perspective well-being is highly subjective as everybody evaluates their lives and their living circumstances in their own way of thinking with socially and culturally circumscribed criteria.

In literature conceptions of well-being appears to follow two main approaches which are the hedonic and eudaimonic traditions. The hedonic

DOI: 10.4018/978-1-7998-4231-6.ch001

tradition gives importance to constructs such as happiness, positive affect, low negative affect, and satisfaction with life (Bradburn, 1969; Diener, 1984; Kahneman, Diener, & Schwarz, 1999; Lyubomirsky & Lepper, 1999). On the other hand the eudaimonic tradition concentrates on the positive psychological functioning and human development (Rogers, 1961; Ryff, 1989; Waterman, 1993). However, despite the differences in approach, most researchers now believe that well-being is a multi-dimensional construct (e.g., Diener, 2009; Michaelson, Abdallah, Steuer, Thompson, & Marks, 2009; Stiglitz, Sen, & Fitoussi 2009). Consequently, the diversity of dimensions has created a confusing and contradictory research base (Pollard and Lee, 2003).

Well-being is related to many factors including everything from physical health, psychological state, level of independence, family, education, wealth, religious beliefs, a sense of optimism, local services and transport, employment, social relationships, housing and the environment (Barcaccia, Esposito, Matarese and others, 2013).

According to Deci and Ryan (2008), subjective well-being is most often interpreted to mean experiencing a high level of positive affect, a low level of negative affect, and a high degree of satisfaction with one's life. According to Bradburn (1969), as one of the first researchers in the field, an individual will be high in psychological well-being in the degree to which he has an excess of positive over negative affect and will be low in well-being in the degree to which negative affect predominates over positive. Similarly, Brey (2015) defines well-being as a state of persons which designates that they are happy or flourishing and that their life is going well for them.

The second major view considers well-being to consist of more than just happiness, suggesting that people's reports of being happy does not necessarily mean that they are psychologically well. This second perspective referred to as eudaimonia (Waterman, 1993) can be traced back to Aristotle and is concerned with living well or actualizing one's human potentials (Deci and Ryan, 2008). There are many researchers that define well-being with an eudaimonic approach. Shah and Marks (2004), argue that well-being is more than just happiness but it means developing as a person, being fulfilled, and making a contribution to the community. Psychologists Riff and Singer (2006) define six key dimensions for psychological well-being as self-acceptance, positive relations with others, autonomy, environmental mastery, purpose in life, and personal growth. According to Naci and Ioannidis (2015), well-being includes choices and activities aimed at achieving physical vitality, mental alacrity, social satisfaction, a sense of accomplishment, and personal fulfillment.

The Well-Being Concept

Often quality of life and well-being are used as similar concepts. Desmet and Pohlmeyer define well-being as a broad concept that represents an individual's overall quality of life. Shin and Johnson (1978) gives a more specific explanation stating that well-being is a global assessment of a person's quality of life according to his own chosen criteria. World health Organization (1997) defines quality of life as an individual's perception of their position in life in the context of the culture and value systems in which they live and in relation to their goals, expectations, standards and concerns. According to Dodge and others (2012), 'quality of life' is used interchangeably with 'well-being' in a variety of disciplines but quality of life appears to be a dimension of well-being rather than an all-embracing definition. Quality of life is related to the qualities that we experience during our life and it is mostly used related to health. On the other hand wellbeing is related to the outcomes of our interaction with the world. In this sense quality of life can contribute to our wellbeing but wellbeing is not only

Another difference between the concepts of wellbeing and quality of life is about the relative temporality of wellbeing. Well-being is often expressed with reference to a specific and temporary situation that could evolve in relatively short time. It is then an inner and subjective state, affected by comfort and quality of life level, but referred to a certain time, to a certain situation, to a certain physical environment or to a certain activity. This is why wellbeing and space are directly in relation to each other as wellbeing can be easily affected by the spatial quality of the spaces we live in. Well-being is felt subjectively, and depends on culture and age, but it has to meet also several "objective" requirements. But of course like quality of life also well-being cannot be ensured by architectural spaces only.

Going further to the relation between well-being and environment, we come across research that focuses on biological aspects. Biologist Stephen Boyden (1971) distinguishes between "survival needs" and "well-being" needs. Survival needs deal with aspects of the environment that directly affect human health, such as clear air and water, lack of pathogens or toxins, and opportunity for rest and sleep. Well-being needs, on the other hand, are more indirect in their locus of impact. These needs affect overall health through their relationship to fulfillment, quality of life, and psychological health. Where failure to satisfy survival needs may lead to serious illness or death, failure to satisfy the well-being needs produces the "grey life" of psychosocial maladjustment and stress related illnesses. Taken as a whole, the research by Boyden and others identifies well being needs that should be addressed in building design:

- Opportunity to engage in spontaneous social encounters.
- Opportunity for relaxation and psychological restoration.
- Opportunity for privacy and for movement between interaction and solitude, as desired.
- Opportunity for learning and information sharing.
- Opportunity for connection to the natural environment.
- Opportunity for regular exercise.
- Sound levels not much above or below that of nature.
- Meaningful change and sensory variability.
- An interesting visual environment with aesthetic integrity.
- Sense of social equity and respect.
- Ability to maintain and control personal comfort.
- Making sense of the environment.

As well-being is dependent on many different factors, it's never stable. In their book "Understanding Happiness" Headey and Wearing (1992) discuss that personality, life events, well-being and psychological distress are in dynamic equilibrium. That is over time they tend to remain in the same balance relative to each other. Stable personality performs a key equilibrating function, tending always to pull all components back to their equilibrium level. Lyubomirsky, Sheldon, and Schkade (2005) discuss that three major factors contribute to people's levels of well-being: 1) their happiness set point (i.e., the genetically determined stable level of happiness, which has been shown to account for approximately 50% of the variance in individual differences in well-being), 2) their life circumstances (e.g., factors such as income, marital status, or religiosity, which are typically found to account for roughly 10% of individual differences in well-being), and 3) positive cognitive, behavioral, and goal-based activities (which have the potential to account for a significant portion – up to 40% – of individual differences in well-being). They also discuss that sustainable changes in well-being are possible to achieve especially through intentional activity that fit one's values and interests and avoiding basing happiness on particular circumstances and objects.

In a recent article Boyden (2016), argues that the health needs of any particular kind of animal, depends on the conditions under which it evolved, because we can be sure that these conditions are capable of providing all the essential ingredients for maintaining and promoting health in that species. Taking our knowledge of the conditions of life of hunter–gatherers as a starting point, we can put together a working list of physical and psychosocial conditions likely to promote health and well-being in our species (Table

1). They are referred to as universal health needs because they apply to all members of the human species wherever or whenever they may be living (Boyden, 2016).

Table 1. Physical and psychosocial conditions likely to promote health and well-being depending on the theory that the conditions of life of hunter–gatherers are the best conditions

Physical Needs	Psychosocial Needs
• Clean air (not contaminated with hydrocarbons, sulfur oxides, etc.). • A natural diet, that is, foods containing the full range of nutritional requirements, as provided by a diverse range of different plant foods and a small amount of cooked lean meat and offal; foods devoid of noxious contaminants or additives. • Clean water (free of contamination with chemicals or pathogenic organisms). • Electromagnetic radiation within the natural range. • Minimal contact with parasites and pathogenic organisms. • Natural contact with non-pathogenic microorganisms in the environment. • Adequate protection from extremes of climate. • Noise levels within the natural range. • A pattern of physical exercise that involves some short periods of vigorous muscular work and longer periods of medium and varied muscular work, but also frequent periods of rest.	• An emotional support network, providing a framework for care-giving and care-receiving behavior. • The experience of conviviality. • Levels of sensory stimulation neither much lower nor much higher than those of the natural habitat.• Variety in daily experience. • Opportunities and incentives for creative behavior and practicing manual skills.• Opportunities and incentives for active involvement in recreational activities (e.g., dancing, making music). • An environment and lifestyle conducive to a sense of belonging, challenge, self-fulfillment, comradeship, love, purpose, and personal involvement in daily activities. • An environment and lifestyle that do not promote a sense of alienation, anomie, deprivation, boredom, loneliness, or chronic frustration.

From a biologist point of view, Boyden argues that cultural evolution has led to numerous instances of cultural maladaptation resulting in ill health in human populations throughout the history of civilization. Many forms of ill health in our society today are the result of evodeviations including most cases of lung cancer, coronary heart disease, obesity, and probably much mental depression. Boyden underlines the survival of the civilization is dependent on the passage to biosensitive society which is in tune with, sensitive to, and respectful of the processes of life that underpin our existence. A biosensitive society will promote health and well-being in all sections of the human population and in the ecosystems of the biosphere. The transition to a biosensitive society will involve sweeping changes in the intensity and nature of human activities. It will require a major scaling down of resource and energy use and the eradication of all practices and technologies that interfere with or threaten human or ecosystem health. It will require revolutionary changes in societal arrangements, including the economic system and the occupational structure of society. Boyden calls this cultural transformation a "cultural renaissance"or perhaps "biorenaissance" because many hunter–gatherer and early farming cultures in the past were characterized by a profound respect for the living world and an understanding that humans are part of nature and completely dependent on other life forms for their well-being and survival.

After evaluating different understandings of wellbeing, it is important to establish the concept in relation to spatial design. In this perspective, wellbeing can be understood as a combination of hedonic and eudaimonic traditions, as the positive state of mind and body in relation to the complex system of interconnected components that built up living. Living environments are essential to prepare the basis for all the qualities that enrich our lives both physically and emotionally. In this sense the concept of wellbeing related to space goes beyond quality of life embracing all aspects of interaction between human and space. Our perceptions, emotions, experiences and their outcomes play an important role on our happiness and satisfaction. A real concept of wellbeing cannot be established without a broader and deeper understanding of human space interaction. Therefore this chapter and this book aims to discuss the spatial dimension of wellbeing mostly overlooked although it is an essential component that can be highly controlled and interpreted in a way that can significantly enhance wellbeing.

WHAT CAN DESIGN DO FOR WELL-BEING?

The role of design and technology is not studied efficiently by social sciences. 17th century philosopher René Descartes held that the technological application of science would yield an unlimited number of devices that would allow people to effortlessly enjoy all benefits that the earth could offer them (Descartes,1637). The optimistic vision of technology of the Enlightenment is still present in contemporary society, in which technology is frequently conceived of as an instrument of social and economic progress that makes people's lives better (Brey, 2015). Brey (2015), argues that on the other hand, 20thcentury philosophy of technology focused on negative implications of technology for well-being and it argued that technology had gone out of control, and that humanity was made subservient by it. He then discusses that it is possible to design for well-being as it may be possible to design artifacts in such a way that a certain desired consequence is bolstered or an undesirable consequence is avoided and increased well-being is a possible consequence of the use of a technological artifact.

According to Larsson and others (2005) design for well-being is an attempt to go beyond the traditional scope of assistive technology by helping people make a transformation from an actual state of being to a desired state of being – regardless of ability level. They define 3 main goals including designing innovative products for increased well-being, shaping the future of globally distributed collaboration and educating the product innovators of tomorrow. Similarly, Miller and Kälviäinen (2006) discuss that well-being results from what people do and decide, rather than what is supplied for them and suggest that it is possible to devise well-being-promoting design heuristics. In this sense, well-being can be supported by encouraging and enabling effective action, prediction and control, satisfying social interaction, and mindfulness, physical involvement and enjoyment.

When we consider design in general we think about consumer goods. Therefore design objects are to be designed produced and sold. Today the products that facilitate our lives require to some degree material wealth. Desmet and Pohlmeyer discuss that material wealth can support individuals in their pursuit of happiness, but it is not a direct source of happiness in itself. This change of focus from material to more personal values also aligns with the findings of psychologists examining the conditions for human flourishing (Seligman, 2011). Numerous studies have confirmed that it is not personal resources that make a person happy, but rather how those resources are exploited (Biswas-Diener, 2008). This idea of material wealth as a resource for happiness opens up a different perspective for design, given that consumer products are also resources. Desmet and Pohlmeyer in their article give examples of design objects that can enable, stimulate, and inspire engagement in meaningful activities. They introduce the positive design framework that includes three main components of subjective well-being which are defined as pleasure, personal significance and virtue. This requires the designer to formulate a vision of how the new design will evoke positive affect, how it will stimulate and enable people in pursuing their personal goals, and how it will support them in being a morally good person.

If well-being is understood as actualizing one's human potentials so that he can make the appropriate contribution to the society, the role of design can be defined as creating the right tools with which a person can develop himself and find the right ways to perform his actions for his specific contribution. Today we are surrounded by designed objects, which help us to perform our actions from our domestic environment to working places, schools and public spaces. These objects have the potential to change our interaction with our surroundings in physical, social and psychological levels. The capability approach claims that in order to attain a state of well-being, people must be in possessing of a set of basic capabilities, which are opportunities to do things or be things that are of value to them. Oosterlaken sees technological development and industrial design as an expansion of human capabilities, and thinks that the details in design should be considered in this perspective (Oosterlaken 2009). Capabilities theorists promote the idea that working to account for and advance human capability strengthens governance, not just at the level of macroeconomic measures but also in terms of civic engagement and citizenship (Dong, 2008).

If we look at the concept of well-being from the perspective of the hedonic tradition, design can promote pleasurable experiences and block undesired experiences. So every design object that involves us in desired activities can contribute to our well-being. Jordan (2000) claims that designers should not just design for functionality and usability, but also for pleasure: products should be pleasurable to use. He distinguishes four types of pleasure, which he claims motivate humans to use products. These are physio-pleasure defines as bodily pleasure deriving from the sense organs; psycho-pleasure as pleasure deriving from cognitive and emotional reactions; socio-pleasure as pleasure arising from one's relationship with other people or society as a whole and ideo-pleasure defined as pleasure arising from people's cultural, aesthetic, moral values, tastes and personal aspirations.

Psychologist Csikszentmihalyi in his book Flow (1990), claims that when people engage in activity, they may reach a state of flow, a feeling of complete and energized focus in an activity, with a high level of enjoyment and fulfillment. In flow, there is a good balance between challenge and skill, so that the task at hand is neither too challenging nor not challenging enough, and someone's skills are experienced as making a good fit with the task at hand.

Brey (2015) discusses that, on a desire-fulfillment conception, design focuses on satisfying desires of users and other stakeholders. Within a simple desire-fulfillment approach, this would require investigations into what people desire and then creating designs to help these desires come true (and that avoids side-effects that are not found desirable). Design is more difficult within an informed desire-fulfillment approach, since it requires determination of the hypothetical desires that people would have if they were properly informed, which is a more speculative endeavor. On an objective list conception, finally, designs should be such that they help bring about the acquisition of goods on the objective list. For example, they should support the development and maintenance or exercise of friendships, autonomy, and practical wisdom (Brey, 2015).

Next, the aggregation problem is described by Van de Poel (2012) as the problem of how the well-being of different people can be aggregated into an overall measure of well-being. The aggregation problem exists because different people will often have different and possibly even conflicting or incommensurable conceptions of well-being, and this raises the question of how different well-being values of different stakeholders can be added up and combined into one measure of well-being and one design (Brey, 2015).

THE SPATIAL DIMENTION OF WELL-BEING

Well-being is a very ephemeral condition that changes from culture to culture, person to person together with time and space. When we talk about well-being and space its important to note that space cannot solve all the problems of human beings but it can always contribute to this positive state of body and mind. For example a city that is donated with attractive sport centers, parks and green areas for recreation will surely attract people giving them the possibility to be in relation to nature and perform sports which will surely lead to a positive state of mind and body. Regarding interiors, the continuity of spatial experience from urban to interior scale, the relation between the natural environment can be important factors that improve well-being. The Japanese house for instance is well known for the strong relation between inside and outside in a way one can perceive the natural environment as an extension of the interior.

Researchers from different fields have underlined the effect of the environment on well-being. More than the built environment the positive effect of the natural environment has been considered to be very important. We experience the healing affect of nature and natural environment in our everyday lives. According to Shah and Peck (2005), nature's role in what we might call our 'spiritual well-being' is unique, and provides a range of benefits including making us more reflective, contributing to our personal growth, inspiring awe, and evoking a feeling of wholeness and belonging. On the other hand, in every stage of our lives, we perform most of our activities indoors, generally not in close relation with the natural environment. Therefore the relation between the built environment and well-being arises as an important issue to be deeply studied.

In the book “A philosophy of interior design”, Abercrombie (1990) states that interiors have a power over us that façades can never have. This is not due to the commonly observed fact that we spend most of our time indoors; it is due instead to the fact that interiors surround us. We do not merely pass them on the street; we inhabit them. When we enter a building, we cease being merely its observer; we become its content. We never fully know a building until we enter it.

Interiors are the spaces that are composed of surfaces and objects that are nearest to the human scale. Most of these objects have very close relationship with human body and perception. According to Cecla (2008), objects are never as they are, they have an evocative power and they are often tools to create symbolic ties with other people. For Weinthal (2011), architecture, body and objects all meet at the scale of the interior, where external forces find their way into the representation of the architectural interior through comparisons of body and world, objective and subjective, natural and artificial. Weinthal (2011) also states that the interior begins with the elements that are closest to the body, forming concentric and more complex layers. As it progresses from the body into spaces, where larger scales are accommodated, the elements keep their relatedness to the body and emotion. Working with interiors means working at full scale in order to determine all aspects of objects surrounding the human body. Every object in interior space has a potential to communicate meaning through human experiences. The material culture of fashion, furniture and interior design narratives identity and reveals the traces of inhabitation, or “being in the world”. “Being in the world” is defined as the most important human mode of perception by Maurice Merleau-Ponty (1945). We accord purpose and meaning to the world around us through our bodies. Shouwenberg (2011) argues that before we can give a name at conscious level to what we see, we have already accumulated by purely somatic means our knowledge about space, the objects in it and our relation to those objects.

The relation between human and built environment is systematically handled in especially work environments. The discipline of macroergonomics is particularly important as it tries to apply scientific knowledge about human capabilities, limitations, and other characteristics in order to improve the human condition, including health, safety, comfort, productivity, and quality of life (Human Factors and Ergonomics Society, 1998). Conceptually, macroergonomics may be defined as a top-down sociotechnical systems approach to the design of work systems, and the carry-through of the overall work system design to the design of the human-job, human-machine, and human-software interfaces (Hendrick, 1997; Hendrick & Kleiner, 2001). As a perspective, macroergonomics provides the ergonomist with an appreciation for the larger system. The goal of macroergonomics is to optimize the work system's design in terms of its sociotechnical system characteristics, and then carry the characteristics of the overall work system design down through to the design of individual jobs and human-machine and human-software interfaces to ensure a fully harmonized work system (Henrick, Kleiner, 2002). In this sense the approach of macroergonomics can be regarded as an important contribution to wellbeing research. On the other hand wellbeing related to space has many other different dimensions other than ergonomic considerations, which will be discussed in Chapter 5 in the Wellbeing Framework for interiors.

INTERIOR SPACE TO PROMOTE WELL-BEING

According to the International Federation of Interior Architects and Designers, one of the most important responsibilities of interior architects and designers is to design interior spaces in order to enhance the quality of life and protect the health, safety, well-being of the public. Smith, Metcalfe and Lommerse (2012) discuss that socially responsive interior architecture is an essential factor in providing the condition for the well-being of individuals and communities. Recently, Petermans and Pohlmeyer discussed the potential of interior architecture to contribute to subjective well-being and pointed to the value of applying a human-centered perspective in the design process.

Well-being in interior space depends on many factors. In Chapter 5 The well-being framework for interiors tries to identify the requirements for well-being in interiors underlining the contextual, functional, ergonomic, social, psychological, sensorial and aesthetic dimensions of well-being. But the fulfillment of requirements sets the base for the issues to be handled in interior design. Every design problem is unique and therefore it is difficult to talk about a strict framework for design. On the other hand, we can talk about qualities that can be considered as promoters of well-being.

Flexibility

Private or public, historic or contemporary, all interiors we live in are subject to continuous change because of the shifting dynamics of everyday life. Interiors do not only change because of changing needs and expectations but also because of the effects of fashion. According to Potvin (2010), interior design marks out a liminal space and enterprise dubiously situated somewhere between fashion and architecture. Fashion is not simply what is displayed on the runway in a magazine or a department store window; it permeates every aspect of the quotidian. Clothing, architecture and interiors all respond to fashion. Clothing and interiors register these changes more quickly than architecture. Commercial interiors change in order to demonstrate that the business they represent is up to date. Public interiors change because of public expectations, needs and tastes. Private interiors change to show that their inhabitants are able to catch up with the changing world around them (Saglar Onay, 2015). The speed of technological change is reflected on the objects people use and objects are continuously disposed and replaced for the desire to stay current or up to date. Architectural construction has higher costs, both in terms of funds and its marks on the landscape compared to fashion and interiors that thrive on the immediacy of change. Weinthal (2011) argues that because of their changing character, interiors are sometimes considered subordinate to architecture, as if lack of permanence consigns them to a superfluous existence. This ephemeral nature contributes to the identity of interior space as frivolous, excessive and without content. Today's interiors are subject to change more than any other period of time and temporality becomes one of the key factors in their definition.

Regarding the mission of interior space as the follower of life, it would be impossible to neglect the role of interior design on our well-being. Interiors are the most dynamic components of the built environment with their continuous tendency to change. The temporary character of interior design becomes an advantage while we think about its capability to support human activity as human needs and expectations can change very rapidly. Although regarded as ephemeral or temporary compared to architecture, interior design has the potential to change people's way of living. In this sense, flexibility almost becomes a requirement for interiors in order to catch up with the needs of the society.

Flexibility is a term that has started to be discussed at the beginning of the 20th century and it is closely related to the functional setup of buildings. Early flexibility concepts were based on donating space with necessary equipment so that it can serve for a number of predefined functions. On the other hand flexibility strategies that have begun to develop after 1950's proposed to start with a new consciousness that all the probable ways of using space can not be predefined beforehand. While the first understanding proposed that everything could be determined by the designer the second had an open-ended structure that embraced uncertainty (Saglar Onay, Garip, 2016).

The term flexibility may refer to various meanings in architecture and interior architecture. The understanding of flexibility may also differ according to cultural factors but it is a feature related to space. It may be considered in various ways, related with temporal variables, functional properties, changing needs of users, different cultural characteristics, different user types, etc. The degree of flexibility is determined in two ways. First the in-built opportunity for adaptability, defined as 'capable of different social uses', and second the opportunity for flexibility, defined as 'capable of different physical arrangements' (Schneider and Till, 2005).

The most crucial point with flexibility is to consider all dimensions by means of analyzing users and their relationships with the environment. Niit (1993) notes that while analyzing housing units as socio-physical systems, basic parameters of "spaces", "relations" and "activities" should be distinguished by phenomenological or other approaches in order to constitute the activity patterns of family members. This may ensure evaluation of various profiles including all family members and their relations with time, space and networks. Schneider and Till note that (2005a, 2005b, 2007), flexibility as accommodating change in housing, addresses a number of issues related with the current and future needs of the users. Firstly, it offers variety in the architectural layout of the units. Secondly, it includes adjustability and adaptability of housing units over time. And finally, it allows buildings to accommodate new functions. In order to provide flexibility, architects should consider the possible future needs of users during the design process; in other words, "long-term thinking" in the design process is required. Albostan (2009) discusses that "flexible housing" refers to the capability of responding to individual wishes and demands over time and offering every user a minimum but efficient basic quality at both base building level and infill or built-in level. Moreover, the changes in the family structure and life cycle of the family members also necessitate flexible solutions.

The concept of flexibility in interior design can be interpreted in different ways. First interiors can be designed in such a way that users can manipulate the aspects that have effect on his/her well-being. Because we know that conditions leading to well-being are not the same for even the members of the same family although they usually have similar cultural backgrounds. These changes that necessitate different spatial solutions need to be carefully analysed by the designer in order to understand the corresponding spatial requirements. At this stage the temporal needs and activities need to be addressed. These may include needs representing different periodic cycles and non-periodic ones. For example a family may accept overnight guests which may necessitate the conversion of the functional layout of a room and this fact can become a very important input that will help to determine the overall layout, spatial components and their relations according to multiple needs. So, multi-functionality arises as an issue to be considered for flexible design. The research project conducted by Saglar Onay, Garip and Garip in Istanbul Basıbuyuk Mass Housing units can be a good example for this. The research based project aims to improve the interior spaces of mass housing projects in accordance with different user needs and to provide solutions that will increase the flexibility of interior spaces. First of all user needs and their temporal changes were analysed by open ended interviews and observations trying to understand the periodic and non periodic activities both dependent and independent from the current interior layout. While interviews helped to understand user's way of living, observations were important to understand the current situation. According to the results obtained from the observations and interviews that were made during the field study, the concept of flexibility was handled in two different scales in the process of interior design. On the upper scale, it was seen that there were different types of household and family typologies in the housing site. For this reason, it was aimed that the proposed system would be able to create scenarios that could be integrated into the cases where the number of people using the house was different. On the lower scale, it was envisaged that the interiors of the houses could work as a flexible system in itself in order to be used in different times of the day and in the case of increasing number of people within home. The survey results showed that the users performed different activities at home in different times of the day. Along with this, the increase in the number of users together with the guests coming home created additional needs within the spaces. Other than this the potentials and limitations of interiors were carefully analysed. Then the main design variables for the flexible use of space were determined as number of family members, presence of guests, activity sets and activity space relationships. The flexible design setup aimed to allow not only the use of interior spaces for different activities, but also the identification of different spatial variations for each activity. This also makes it easier for users to customize the places they live in and to rearrange them according to the number and preferences of the users (Saglar Onay, Garip, Garip, 2019).

Flexibility in interior design can also be interpreted as interior's potential to support creative activity. Recently, neuroscientists are learning that creativity has been a survival function in the evolution of the human brain suggesting that opportunities for creative activity are also very important for human well-being (IOM, 2007). Creative activities are not limited to common conceptualizations of art, such as painting or music (Ivcevic, 2007) but they can be diffused in a variety of daily activities, such as doing crafts, decorating a living environment, solving everyday problems, using humour, and structuring leisure time activities (Murdock &Ganim, 1993; Runco & Bahleda, 1986; Torrance, 1988; Eschleman, Madsen, Alarcon, Barelka, 2014). Creative activity within a single domain is believed to provide resources that enrich one's life, increase personal well-being, and improve performance in other life domains (Baer, 2012;Ivcevic, 2007; Richards, 1990). Evaluating the importance of creative activity, it can be assumed that interior design needs to know where to stop leaving the control to the user donating him by the necessary tools to fulfil his needs. Creative activity is often characterized by producing a unique outcome (Amabile, 1983), and it is likely to result in greater perceived control over outcomes (Eschleman, Madsen, Alarcon, Barelka, 2014). Regarding the issue of control, a great deal of research has shown that, for diverse groups and situations, sense of control is an important factor influencing stress levels and well-being (Steptoe, Appels, 1989; Ulrich, 1991). Supporting these views, Miller and Kälviäinen (2006), discuss that designing for well-being is to design the (metaphorical) space that enables people to 'do well-being' for themselves.

Sense of Place

In his 'Poetics of Space' Bachelard (1958), argues that the relationship between space and personality is so intimate that to understand oneself a topoanalysis, the exploration of self-identity through place, might yield more fruitful insight than psychoanalysis. There are many dimensions of meanings ascribed to place: symbolic, emotional, cultural, political and biological. As with other members of the biosphere, humans display marked patterns of territoriality (Buttimer, Seamon, 1980). On the other hand, the existence of values attached to place are usually taken for granted and people get to understand this only when they tend to be threatened.

Sense of place is increasingly recognized as key to human well-being in social-ecological systems (Poe, Donatuto, Satterfield, 2016). It is in relation to those characteristics that make a place special or unique, as well as to those that foster a sense of authentic human attachment and belonging (Casey, 2001).

Place attachment is also defined as a feeling of comfort, which the individual is benefited from in his or her living place (Brown, Perkins, & Brown, 2003). According to Wang et al. (2016), improving the place attachment leads to the promotion of social well-being and social cognitive function. Rollero & Piccoli (2010), also conducted a study on 443 students at the University of Turin in Italy to assess the effect of attachment to place on social well-being and found that attachment to place has considerable effects on social well-being (Rollero & De Piccoli, 2010). So we can say that correct relationships with context increase place attachment, which is absolutely considered as a key concept in well-being.

Although the issue of place attachment seems as an issue that needs to be handled in urban and architectural scales, the role of interiors is also very significant. Sense of place in living environments is about context as well as the identity or unique characteristics of a building or space that create meaning for an occupant. In this sense, the values and characteristics that identify place need to be represented in all scales subject to human experience. Regarding interior design practices today, we come across many examples that aim to create a completely contrasting interior environment with the surroundings. As the interior designer has the potential to evoke the feelings associated with the meanings he or she wants to communicate, he should always consider contextual ties as one of the most important factors in determining spatial identity.

Contact with Nature

Contact with nature is another issue to be considered in well-being and space studies. A growing body of knowledge supports the role of contact with nature in human well-being. Kellert (2016) notes that contact with nature has been found to enhance recovery from illness, that people living in proximity to open spaces report fewer health and social problems as well as superior quality of life and a stronger sense of place, that office settings with natural environmental features improve worker performance and motivation (Freeman, 2017). These studies provide scientific support for the assumption that contact with nature is critical to human well-being (Saglar Onay, Minucciani, 2018). It is very important that more attention is given to social and environmental contexts and their relations to well-being in interpersonal and social levels in future studies.

Kamitsis and Francis (2013) argue that engagement with nature, through both direct sensory exposure and a sense of connectedness, has been shown to have a positive effect on psychological health through their research; although the mechanisms mediating these effects remain obscure. Capaldi and others (2015), discuss that there are two ways of being involved in nature, which are nature contact and nature connectedness. Nature contact involves interacting with the natural world, such as being immersed in a natural environment, being around natural elements indoors (e.g., plants), or being exposed to virtual representations of nature (e.g., viewing photographs or videos of natural landscapes). Nature connectedness, however, refers to one's subjective sense of connection with the natural world. Although typically conceived of as an individual difference (Tam, 2013), nature connectedness can also be measured as a state; it can fluctuate in the short-term depending on situational context. Individuals who are more connected to nature spend more time outdoors (Mayer & Frantz, 2004; Nisbet, Zelenski, & Murphy, 2009; Tam, 2013), and nature contact often increases momentary feelings of connectedness (Mayer, Frantz, Bruehlman-Senecal, & Dolliver, 2009; Nisbet & Zelenski, 2011; Capaldi, Passmore, Nisbet, Zelenski, Dopko, 2015, 2015). Contact with nature and nature relatedness has been linked with several indicators of eudaimonic well-being. For example, feeling that one's life is meaningful correlates positively with nature connectedness (e.g., Cervinka, Roderer, & Hefler, 2012; Nisbet, Zelenski, & Murphy, 2011: Capaldi, Passmore, Nisbet, Zelenski, Dopko, 2015).

Natural environments seem particularly restorative: they provide an opportunity to get away, contain fascinating rich stimuli that effortlessly engage our involuntary attention, and allow us to act without the need to constantly monitor our behavior (Kaplan & Kaplan, 1989). Numerous empirical studies report improvements in concentration, directed attention (Berman, Jonides, & Kaplan, 2008; Berto, 2005; Hartig, Evans, Jamner, Davis, & Gärling, 2003; Van den Berg et al., 2003), and emotional functioning (discussed below) after contact with nature (Capaldi, Passmore, Nisbet, Zelenski, Dopko, 2015). A survey of 3,000 Finnish participants found that those who reported spending more of their leisure time engaged in nature-based recreational activities had higher emotional well-being than those who reported spending less of their free time in nature (Korpela, Borodulin, Neuvonen, Paronen, & Tyrväinen, 2014; Capaldi, Passmore, Nisbet, Zelenski, Dopko, 2015).

Empirical evidence links exposure to nature with increased autonomy, the ability to freely choose one's actions—an important part of eudaimonic well-being. Research from the outdoor education and experiential learning literature describes similar benefits of wilderness and nature immersion experiences on autonomy and other measures of psychological well-being like personal growth, self-esteem, self-regulation, and social competency (e.g., Norton & Watt, 2014; Passarelli, Hall, & Anderson, 2010; Ray & Jakubec, 2014, Capaldi, Passmore, Nisbet, Zelenski, Dopko, 2015)

Preliminary evidence suggests that repeated contact with nature produces larger increases in well-being than other interventions commonly cited in the positive psychology literature (Passmore & Howell, 2014). Regarding living environments contact with nature can be evaluated in urban, architectural and interior levels. This necessitates to handle the issue in an interdisciplinary way including professionals working at different scales. The continuation of spatial experience needs to be accompanied by nature in a natural way. Research shows that the amount of neighborhood green space is correlated with stronger social ties reported among neighbors and greater prosocial activity in the neighborhood (Kuo, 2003; Kuo, Sullivan, Coley, & Brunson, 1998; Sommer, 2003; Sullivan, Kuo, & Depooter, 2004; Capaldi, Passmore, Nisbet, Zelenski, Dopko, 2015). Consequently Strong social ties increase the psychological well-being of inhabitants.

There are differing design approaches that search for the best methods and practices to involve nature and natural aspects in the design process such as biomimicry and biophilic design, which will be discussed in Chapter 4 titled different design approaches to well-being. The biophilia hypothesis posits that our ancestors' well-being and survival depended on connecting with nature (i.e., for finding food and water, navigating, and predicting time or future weather conditions, etc.; Kellert & Wilson, 1993; Capaldi, Passmore, Nisbet, Zelenski, Dopko, 2015). While biomimicry focus on practical solutions inspired by natural forms, processes and systems, biophilia describes humans' connection with nature and biophilic design is replicating experiences of nature in design to reinforce that connection. Biomimicry's goal is in achieving better performance, while biophilic design aims to improve health and well-being. Biomimicry is more heavily used in technology and product development, while biophilia concerns more directly to interior design, as well as to architecture and urban design (Minucciani, Saglar Onay, 2019).

Regarding the relationship between interior space and nature in history, we can see the industrial revolution as a breaking point. In many countries, the increase in urban density and poor urban design strategies have let to an isolation from nature. And today architects and designers try to pump natural aspects into existing living environments that suffer from the lack of nature. Although a positive tendency, these interventions mostly remain at a superficial level as a real understanding of connectedness to nature requires a holistic environmental approach. The qualities that can be observed in vernacular living environments represent strong ties with nature supporting their daily activities, social interactions and cultural rituals which can be regarded as aspects that have very significant benefits on well-being. These aspects will be discussed in Chapter 9_ Cultural Components of Well-being and Vernacular Examples.

FUTURE RESEARCH DIRECTIONS

The lack of research studies in the intersection of social sciences and design theory/ practice necessitates an interdisciplinary approach. Therefore, future research on well-being and space needs to be structured by the participation of researchers from sociology, psychology, environmental psychology, medicine, neurosciences and all design disciplines including urban design, landscape design, architecture, industrial and interior design.

CONCLUSION

Through this chapter, the connections between well-being and interior space have been discussed by evaluating recent research in the fields of psychology, sociology, ecology and design. Researchers from different fields have underlined the effect of the environment on well-being but the role of interior design have been addressed in a very limited number of studies. This fact is difficult to understand as the interior environment is composed of surfaces and objects that are nearest to the human scale.

Well-being research in general is dominated by two different traditions. The hedonic tradition gives importance to constructs such as happiness, positive affect, low negative affect, and satisfaction with life while the eudaimonic tradition concentrates on the positive psychological functioning and human development. By many researchers from different disciplines the concept of wellbeing is considered as sum of both views. Correspondingly in this book wellbeing is evaluated as the positive state of mind and body in relation to all interconnected aspects of life.

Regarding the importance of design from the perspective of both hedonic and eudaimonic traditions, design can create the right tools with which a person can develop himself and find the right ways to perform his actions for his specific contribution and design can also promote pleasurable experiences and block undesired experiences. Designed objects have the potential to change our interaction with our surroundings in physical, social and psychological levels.

Moving to the role of interior design we can say that interiors can provide the right environment in which people can be free to perform the actions they choose in the way they want and they can appreciate pleasurable experiences as an outcome of their interaction with space. In this sense the interior environment needs to be evaluated as a very strong determinant of general well-being. This is also because interiors mark out a domain that is controllable in the first step by the designer and in the second step by the user. This control mechanism needs to be structured in such a way that interior design creates the appropriate stage for living and gives the necessary tools to the user to decide on his well-being himself.

REFERENCES

Abercrombie, S. (1990). *A philosophy of interior design*. Avalon Publishing.

Amabile, T. M. (1983). The social psychology of creativity: A componential conceptualization. *Journal of Personality and Social Psychology*, *45*(2), 357–376. doi:10.1037/0022-3514.45.2.357

Baer, J. (2012). Domain specificity and the limits of creativity theory. *The Journal of Creative Behavior*, *46*(1), 16–29. doi:10.1002/jocb.002

Barcaccia, B., Esposito, G., Matarese, M., Bertolaso, M., Elvira, M., & De Marinis, M. G. (2013). Defining quality of life: A wild-goose chase. *Europe's Journal of Psychology*, *9*(1), 185–203. doi:10.5964/ejop.v9i1.484

Berman, M. G., Jonides, J., & Kaplan, S. (2008). The cognitive benefits of interacting with nature. *Psychological Science*, *19*(12), 1207–1212. doi:10.1111/j.1467-9280.2008.02225.x PMID:19121124

Berto, R. (2005). Exposure to restorative environments helps restore attentional capacity. *Journal of Environmental Psychology*, *25*(3), 249–259. doi:10.1016/j.jenvp.2005.07.001

Biswas-Diener, R. M. (2008). Material wealth and subjective well-being. In M. Eid & R. J. Larsen (Eds.), *The science of subjective well-being* (pp. 307–322). Guilford Press.

Boyden, S. (1971). Biological determinants of optimal health. In D.J.M. Vorster (Ed.), *The human biology of environmental change. Proceedings of a conference held in Blantyre Malawi*. London: International Biology Program.

Boyden, S. (2016). The biohistorical paradigm: The early days of human ecology at the Australian National University. *Human Ecology Review*, *22*(2), 25–46. doi:10.22459/HER.22.02.2016.02

Bradburn, N. (1969). *The structure of psychological well-being*. Aldine.

Brey, P. (2015). Design for the value of human well-being. In J. van den Hoven, P. Vermaas, & I. van de Poel (Eds.), *Handbook of Ethics, Values, and Technological Design. Sources, Theory, Values and Application Domains* (pp. 365–382). Springer. doi:10.1007/978-94-007-6970-0_14

Brown, B., Perkins, D. D., & Brown, G. (2003). Place attachment in a revitalizing neighborhood: Individual and block levels of analysis. *Journal of Environmental Psychology*, *23*(3), 259–271. doi:10.1016/S0272-4944(02)00117-2

Buttiber, A., & Seamon, D. (2015). *The Human Experience of Space and Place*. Routledge. doi:10.4324/9781315684192

Capaldi, C. A., Passmore, H. A., Nisbet, E. K., Zelenski, J. M., & Dopko, R. L. (2015). Flourishing in nature: A review of the benefits of connecting with nature and its application as a well-being intervention. *International Journal of Wellbeing*, *5*(4), 1–16. doi:10.5502/ijw.v5i4.449

Casey, E. S. (2001). Between geography and philosophy: What does it mean to be in place world? *Annals of the Association of American Geographers*, *91*(4), 683–693. doi:10.1111/0004-5608.00266

Cecla, F. L. (2008). *Contro Architettura*. Bollati Boringhieri.

Cervinka, R., Roderer, K., & Hefler, E. (2012). Are nature lovers happy? On various indicators of well-being and connectedness with nature. *Journal of Health Psychology*, *17*(3), 379–388. doi:10.1177/1359105311416873 PMID:21859800

Deci, E. L., & Ryan, R. M. (2008). Hedonia, eudaimonia, And well-being: An introduction. *Journal of Happiness Studies*, *9*(1), 1–11. doi:10.100710902-006-9018-1

Descartes, R. (1994). *Discourse on Method.* Everyman Paperbacks. (Original work published 1637)

Desmet, P. M. A., & Pohlmeyer, A. E. (2013). Positive design: An introduction to design for subjective well-being. *International Journal of Design, 7*(3), 5–19.

Diener, E. (1984). Subjective well-being. *Psychological Bulletin, 95*(3), 542–575. doi:10.1037/0033-2909.95.3.542 PMID:6399758

Diener, E. (2009). The science of well-being: The collected works of Ed Diener. Social indicators research series, 37. Springer Science + Business Media.

Dodge, R., Daly, A., Huyton, J., & Sanders, L. (2012). The challenge of defining well-being. *International Journal of Wellbeing, 2*(3), 222–235. doi:10.5502/ijw.v2i3.4

Dong, A. (2008). The policy of design: A capabilities approach. *Design Issues, 24*(4), 76–87. doi:10.1162/desi.2008.24.4.76

Eschleman, K. J., Madsen, J., Alarcon, G., & Barelka, A. (2014). Benefiting from creative activity: The positive relationships between creative activity, recovery experiences, and performance related outcomes. *Journal of Occupational and Organizational Psychology, 87*(3), 579–598. doi:10.1111/joop.12064

Freeman, K. (2017). *How wellness and well-being went mainstream.* Retrieved from https://www.ambius.com/blog/how-wellness-and-well-being-went-mainstream/

Hartig, T., Evans, G. W., Jamner, L. D., Davis, D. S., & Gärling, T. (2003). Tracking restoration in natural and urban field settings. *Journal of Environmental Psychology, 23*(2), 109–123. doi:10.1016/S0272-4944(02)00109-3

Headey, B., & Wearing, A. J. (1992). *Understanding happiness: A theory of subjective well-being.* Longman Cheshire.

Hendrick, H. W. (1997). Organizational design and macroergonomics. In G. Salvendy (Ed.), *Handbook of human factors and ergonomics* (2nd ed., pp. 594–636). Wiley.

Hendrick, H. W., & Kleiner, B. M. (2001). *Macroergonomics: An introduction to work system design.* Human Factors and Ergonomics Society.

Henrick, H. W., & Kleiner, B. M. (2002). *Macroergonomics: theory, methods, and applications.* Lawrence Erlbaum Associates, Publishers. doi:10.1201/b12477

Human Factors and Ergonomics Society. (1998). HFES strategic plan. *Human Factors and Ergonomics Society Directory and Yearbook, 1998-1999.*

Ivcevic, Z. (2007). Artistic and everyday creativity: An act-frequency approach. *The Journal of Creative Behavior, 41*(4), 271–290. doi:10.1002/j.2162-6057.2007.tb01074.x

Kahneman, D., Diener, E., & Schwarz, N. (Eds.). (1999). *Well-being: The foundations of hedonic psychology.* Russell Sage Foundation.

Kamitsis, I., & Francis, A. J. P. (2013). Spirituality mediates the relationship between engagement with nature and psychological well-being. *Journal of Environmental Psychology, 36*, 136–143. doi:10.1016/j.jenvp.2013.07.013

Kellert, R. (2005). *Building for life: designing and understanding the human-nature connection.* Island Press.

Kellert, S. R., & Wilson, E. O. (1993). *The biophilia hypothesis.* Island Press.

Korpela, K., Borodulin, K., Neuvonen, M., Paronen, O., & Tyrväinen, L. (2014). Analyzing the mediators between nature-based outdoor recreation and emotional well-being. *Journal of Environmental Psychology, 37*, 1–7. doi:10.1016/j.jenvp.2013.11.003

Kuo, F. E. (2003). Social aspects of urban forestry: The role of aboriculture in a healthy social ecology. *Journal of Arboriculture, 29*, 148–155.

Kuo, F. E., Sullivan, W., Coley, R., & Brunson, L. (1998). Fertile ground for community: Inner-city neighbourhood of common spaces. *American Journal of Community Psychology, 26*(6), 823–851. doi:10.1023/A:1022294028903

Larsson, A., Larsson, T., Leifer, L., Van der Loos, M., & Feland, J. (2005). Design for well-being: Innovations for people. *Proceedings of International Conference on Engineering Design ICED 05.*

Lyubomirsky, S., & Lepper, H. S. (1999). A measure of subjective happiness: Preliminary reliability and construct validation. *Social Indicators Research, 46*(2), 137–155. doi:10.1023/A:1006824100041

Lyubomirsky, S., Sheldon, K. M., & Schkade, D. (2005). Pursuing happiness: The architecture of sustainable change. *Review of General Psychology*, *9*(2), 111–131. doi:10.1037/1089-2680.9.2.111

Mayer, F. S., Frantz, C. M., Bruehlman-Senecal, E., & Dolliver, K. (2009). Why is nature beneficial? The role of connectedness to nature. *Environment and Behavior*, *41*(5), 607–643. doi:10.1177/0013916508319745

Merleau-Ponty, M. (1945). *La Phénoménologie de la Perception*. Gallimard.

Michaelson, J., Abdallah, S., Steuer, N., Thompson, S., Marks, N. (2009). *National accounts of well-being: Bringing real wealth onto the balance sheet*. London: Nef (The New Economics Foundation).

Miller, H., & Kälviäinen, M. (2006). Design for Wellbeing. In *Proceedings of Design and Emotion 06 Conference*. Gothenburg, Sweden: Chalmers University.

Minucciani, V., & Saglar Onay, N. (2018). Evaluation of design approaches for well-being in interiors. *Journal of Engineering and Architecture*, *6*(1), 112–122.

Murdock, M. C., & Ganim, R. M. (1993). Creativity and humor: Integration and incongruity. *The Journal of Creative Behavior*, *27*(1), 57–70. doi:10.1002/j.2162-6057.1993.tb01387.x

Naci, H., John, P. A., & Ioannidis, M. D. (2015). Evaluation of wellness determinants and interventions by citizen scientists. *Journal of the American Medical Association*, *314*(2), 121–122. doi:10.1001/jama.2015.6160 PMID:26068643

Nisbet, E. K., Zelenski, J. M., & Murphy, S. A. (2009). The nature relatedness scale: Linking individuals' connection with nature to environmental concern and behavior. *Environment and Behavior*, *41*(5), 715–740. doi:10.1177/0013916508318748

Norton, C. L., & Watt, T. T. (2014). Exploring the impact of a wilderness-based positive youth development program for urban youth. *Journal of Experiential Education*, *37*(4), 335–350. doi:10.1177/1053825913503113

Oosterlaken, I. (2009). Design for development: A capability approach. *Design Issues*, *25*(4), 91–102. doi:10.1162/desi.2009.25.4.91

Passarelli, A., Hall, E., & Anderson, M. (2010). A strengths-based approach to outdoor and adventure education: Possibilities for personal growth. *Journal of Experiential Education, 33*(2), 120–135. doi:10.1177/105382591003300203

Passmore, H.-A., & Howell, A. J. (2014). Nature involvement increases hedonic and eudaimonic well-being: A two-week experimental study. *Ecopsychology, 6*, 148–154.

Petermans, A., & Pohlmeyer, A. E. (2014). Design for subjective well-being in interior architecture. *Proceedings of the 6th Annual Architectural Research Symposium*, 206–218.

Poe, M. R., Donatuto, J., & Satterfield, T. (2016). Sense of place: Human well-being considerations for ecological restoration in Puget Sound. *Coastal Management, 44*(5), 409–426. doi:10.1080/08920753.2016.1208037

Potvin, J. (2010). The velvet masquerade: Fashion, interior design and the furnished body. In A. Myzelev & J. Potvin (Eds.), *Fashion, Interior Design and the Contours of Modern Identity*. Ashgate Publishing.

Ray, H., & Jakubec, S. L. (2014). Nature-based experiences and health of cancer survivors. *Complementary Therapies in Clinical Practice, 20*(4), 188–192. doi:10.1016/j.ctcp.2014.07.005 PMID:25160991

Richards, R. (1990). Everyday creativity, eminent creativity, and health: 'Afterview'; for CRJIssues on creativity and health. *Creativity Research Journal, 3*(4), 300–326. doi:10.1080/10400419009534363

Rogers, C. R. (1961). *On becoming a person*. Houghton Mifflin.

Rollero, C., & De Piccoli, N. (2010). Does place attachment affect social well-being? Revue Européenne de Psychologie Appliquée/European. *Review of Applied Psychology, 60*(4), 233–238. doi:10.1016/j.erap.2010.05.001

Runco, M. A., & Bahleda, M. D. (1986). Implicit theories of artistic, scientific and everyday creativity. *The Journal of Creative Behavior, 20*(2), 93–98. doi:10.1002/j.2162-6057.1986.tb00423.x

Ryff, C. D. (1989). Happiness is everything, or is it? Explorations on the meaning of psychological well-being. *Journal of Personality and Social Psychology, 57*(6), 1069–1081. doi:10.1037/0022-3514.57.6.1069

Ryff, C. D., & Singer, B. H. (2006). Best news yet on the six-factor model of well-being. *Social Science Research, 35*(4), 1103–1119. doi:10.1016/j.ssresearch.2006.01.002

Saglar Onay, N. (2015). Interiors seeking a New Design Language. *The International Journal of Architectonic, Spatial, and Environmental Design, 8*(2), 21–30. doi:10.18848/2325-1662/CGP/v08i02/57904

Seligman, M. E. P. (2011). *Flourish*. Free Press.

Shah, H. & Peck, J. (2015, July). Well-being and the environment. *Nef.*

Shin, D., & Johnson, D. (1978). Avowed happiness as an overall assessment of the quality of life. *Social Indicators Research, 5*(1), 475–492. doi:10.1007/BF00352944

Smith, D., Metcalfe, P., & Lommerse, M. (2012). Interior architecture as an agent for well-being. *Journal of the HEIA, 19*(3).

Sommer, R. (2003). Trees and human identity. In S. Clayton & S. Opotow (Eds.), *Identity and the natural environment: The psychological significance of nature* (pp. 179–204). MIT Press.

Steptoe, A., & Appels, A. (1989). *Stress, Personal Control and Health*. John Wiley.

Stiglitz, J. E., Sen, A., & Fitoussi, J. P. (2009). Report by the commission on the measurement of economic performance and social progress. Academic Press.

Sullivan, W. C., Kuo, F. E., & Depooter, S. F. (2004). The fruit of urban nature: Vital neighborhood space. *Environment and Behavior, 36*(5), 678–700. doi:10.1177/0193841X04264945

Tam, K. P. (2013). Concepts and measures related to connection to nature: Similarities and differences. *Journal of Environmental Psychology, 34*, 64–78. doi:10.1016/j.jenvp.2013.01.004

Torrance, E. P. (1988). The nature of creativity as manifest in its testing. In R. J. Sternberg (Ed.), *The nature of creativity* (pp. 43–75). Cambridge University Press.

Ulrich, R. S. (1991). Effects of Interior Design on Wellness: Theory and Recent Scientific Research. *Journal of Health Care Interior Design, 3*, 97-109.

Van de Poel, I. (2012). Can we design for well-being? In P. Brey, A. Briggle, & E. Spence (Eds.), *The Good Life in a Technological Age* (pp. 295–306). Routledge.

Van den Berg, A. E., Koole, S. L., & van der Wulp, N. Y. (2003). Environmental preference and restoration: (How) are they related? *Journal of Environmental Psychology*, *23*(2), 135–146. doi:10.1016/S0272-4944(02)00111-1

Wang, X., Xue, S., & Kong, F. (2016). Amplitude of low frequency fluctuations during resting state predicts social well-being. *Biological Psychology*, *118*, 161–168. doi:10.1016/j.biopsycho.2016.05.012 PMID:27263835

Waterman, A. S. (1993). Two conceptions of happiness: Contrasts of personal expressiveness (eudaimonia) and hedonic enjoyment. *Journal of Personality and Social Psychology*, *64*(4), 678–691. doi:10.1037/0022-3514.64.4.678

Weinthal, L. (2011). Introduction. In L. Weinthal (Ed.), *Toward a New Interior: An Anthology of Interior Design Theory*. Princeton Architectural Press.

World Health Organization. (1997). *WHOQOL Measuring Quality of Life*. Geneva: World Health Organisation. https://ifiworld.org/about/

KEY TERMS AND DEFINITIONS

Eudaimonia: Aristotle's view of happiness that arises from good works, the greatest life lived to its fullest potential in accord with internal virtue.

Macroergonomics: An approach to the design of work systems, and the carry-through of the overall work system design to the design of the human-job, human-machine, and human-software interfaces.

Chapter 2
A Short History of Well-Being in Interiors

ABSTRACT

Although well-being studies started in the psychological field (1961) when well-being was discussed in terms of "the good life," the concept dates back to ancient times from Aristotle's idea of eudaimonia to the human-centered understanding of the Renaissance. During the last decades, different studies focused on social, physiological, and psychological impacts of the dwelling and the neighborhood, and some even applied neurosciences to architecture. The contribution of studies carried in healthcare contexts has been crucial, as well as experimental results in workplaces, but also biologists, psychologists, and sociologists supported theory and practice, starting from the 1960s. Nevertheless, the specific role of architecture and interior space has been quite under-evaluated. This chapter aims to discuss the foundations of the concept of living well evaluating the understanding of "space for human well-being" in different cultural contexts at different times.

INTRODUCTION

In this chapter the focus will be on the concept of "the good life" rather than the scientific concept of well-being. According to Angner (2011), the appearance of a science of happiness was perhaps first marked by Myerson's 1917 study promoting the discipline of eupathics, focusing on "the well being of the normal". But people's search for the good life has always existed.

DOI: 10.4018/978-1-7998-4231-6.ch002

This search was dependent upon the opportunities, threatens and priorities of that period and it was always the follower of the coeval ways of thinking. Therefore what is considered, as good life at a particular time in a particular place may not correspond to the conception of well-being today. In this sense it is important to understand what the good life meant for people in different times, which aspects were more consistent and which were more ephemeral? Of course it is very difficult to make precise assumptions about how happy or satisfied people were with their lives but it is possible to evaluate the clues they left including the ideas about happiness, ideal life, being good and their reflections on architecture.

To evaluate the historical background of the study of well-being it is necessary to formulate its definition. Contemporary psychologists have drawn on the Aristotelian distinction to suggest that modern well-being research falls into two conceptual camps: hedonics, which focuses on how the person feels about his or her life, and eudaimonics, which focuses on living a life in full accord with one's potentials. (Ryan, Deci, 2001). While the hedonic tradition accentuates constructs such as happiness, positive affect, low negative affect, and satisfaction with life (eg, Bradburn, 1969; Diener, 1984; Kahneman et al 1999; Lyubomirsky & Lepper, 1999); the eudaimonic tradition highlights positive psychological functioning and human development (eg, Rogers, 1961; Ryff, 1989a; 1989b; Waterman, 1993) (Dodge et al, 2012). Also in history we can see the existence of these different points of view some of which also have effect on today's conception of well-being.

SEARCH FOR "THE GOOD LIFE" IN TIMES

In most of the studies that focus on the meaning of well-being and measures to understand it, we come across Aristotle's view of Eudaimonia. Aristotle influenced European thought about 'the good life' for 1500 years. According to Jackson (2013), the recent discourse of 'happiness' and growth of 'positive psychology' are perhaps reworkings for today of an Aristotelian construction of 'well-being' and importantly, 'well-being' in this form continues to conjure a vision of all that is best and desirable for a person. Jackson argues that philosophical 'well-being' is a tool for thinking, an idealised aspiration rather than a real state to be attained or measured. According to Aristotle, euphoria is innately ideal: we seek euphoria for the sake of euphoria. It must be innately independent, needing no other entity. It is man's ideal good in itself (Raftari 2015). Similarly Jackson (2013) discusses that for Aristotle well-being was an ideal, the culmination of a person's idealised journey to 'actualise' all his potential. In this form, it certainly has no opposite - there is the ideal of well-being, and the person's potential for well-being, but no 'ill-being'

In his book Nicomachean ethics, Aristotle famously distinguished hedonism (the life occupied by the search for pleasure) and eudaimonia (happiness that arises from good works). Aristotle's view of eudaimonia, was that the greatest life was the one that was lived to its fullest potential or in accord with some internal virtue. He argues that according to masses or ordinary people happiness is something straightforward and obvious like pleasure, wealth or honour. But in fact Eudaimonia is something that we need to search for according to our internal values. In this sense the way to happiness is different for everyone. In Aristotle's conception of happiness, goods have been classified into three groups: those called external goods, goods of the soul, and goods of the body (Crisp, 2000). For him experiencing pleasure is an aspect of the soul, and each person finds pleasure in that of which he is said to be fond. Goods of the soul are the ones we call most strictly and most especially good, and these actions and activities of the soul we may attribute to the soul. Actions in accordance with virtue are like this, so that they are pleasant to these people as well as in themselves (Crisp, 2000). A happy person will spend all or most of his time in action and contemplation in accordance with virtue. And he will bear changes in fortune in a particularly noble way and altogether gracefully, as one who is genuinely good. He will not be shifted easily from happiness, and not by ordinary misfortunes (Crisp, 2000). McMahon (2004) summarizes this approach of good life by saying 'happiness itself was not a function of feeling, but a function of virtue' That is, happiness is here seen as earned through right action. Even so, for Aristotle, good feelings were an integral part of eudaimonia.

In Aristotle's logic, good is spoken of in the categories of substance as for instance God and intellect; in that of quality like the virtues; in that of quantity as the right amount; in that of relation like the useful; in that of time like the right moment, and in that of place like the right locality (Crisp, 2000). So we can note that Aristotle has already mentioned the importance of place and the environment on the concept of good. So something might be evaluated as good if it is there on the right place together with other categories of substance that affect the end product.

For Aristotle happiness is different from other good as it is complete without qualification. Honour, pleasure, intellect and every virtue are chosen for the sake of happiness on the assumption that through them we can live a life of happiness. But the complete good is through to be self-sufficient which he defines as what makes life worthy of choice and lacking nothing. Happiness is obviously something complete and self-sufficient and in that it is the end of what is done. So, the human good turns to be the activity of the soul in accordance with he appropriate virtue (Crisp, 2000).

Regarding the relationship between individual well-being and the environment he lives in, we come across Plato's *Republic*, according to which in the ideal city, harm to one citizen or part of the city is felt as harm to the rest of the citizens or the city as a whole. It was a form of social and political organization that allowed individuals to maximize their potentialities, serve their fellow citizens, and live in accordance with universal laws and truths. The ideal state according to the Platonic doctrine should be based on three categories of people: artisans, auxiliary militaries and philosopher-kings, each with unique nature and qualities. In a fair state, each class has an obligation to remain true to its nature, destined to engage with real commitments and tasks. Proper management of a soul would open the way to happiness and prosperity (Simeonov, 2011). Plato also introduces the principle of the division of labor, to explain why societies tend to be heterogeneous rather than homogeneous. Nothing could guarantee efficiency better than a social arrangement in which all work was done by those best suited to it (Pappas, 1995). Similarly Socrates argues that each person will be happy engaging in the occupation that suits them best. If the city as a whole is happy, then individuals are happy.

Conception of communal happiness can be seen also during the Roman Period. An important evidence about how an Roman citizen (indeed an emperor) might try to live a Stoic life is the Meditations of Marcus Aurelius (121-180 AD). Stoicism teaches that virtue is the only good for oneself, that vice is the only evil, and that everything else is indifferent so far as one's happiness is concerned. That is to say, only virtue can contribute to our happiness; only vice can contribute to our unhappiness. Poverty, ill-repute, and ill-health are not bad, for their possession does not make us unhappy; wealth, fame and good are not good because their possession does not make us happy (Rachana, 2018).

Marcus Aurelius fleshes out 'acting justly' in terms of acting communally and adds that wherever one lives, one should live as a citizen of the cosmic city. Appeal to the idea that the cosmos is a city allows him to say that we should do well for all humanity, for we each have a citizen's duty to contribute to the welfare of the whole cosmopolis, i.e., to the welfare of all humans as our fellow-citizens. Conversely, anyone who does not contribute to the communal goal (*to koinônikon telos*) is acting seditiously; one may not hate even one man, for this rends the community. Sometimes Marcus goes so far as to identify the good (*agathon*) of a rational creature with community. He says, on the one side, that the perfection, well-being, and stability of the whole depends on what happens to each part. And on the other side, he says that what the nature of the whole brings about is good (*agathon*) for each part, and that what is not hurtful to the city can't be hurtful to its citizen (Rachana, 2018). While Plato uses the limb-body analogy to emphasize the unity of feeling the ideal city achieves, Marcus uses it to emphasize that the citizen is a functional part of the whole city: just as this material making up a limb would not be a limb at all without the body of which it is a part, so too, this human individual would not be what they are without a city of which they are a part (Rachana, 2018).

By the time Christianity began to influence Western philosophy, different conceptions of happiness started to develop. St. Augustine of Hippo (354 – 430 AD) was an early Christian theologian and philosopher who claimed that the fundamental problem of human is misplaced loves. He discusses that people love the wrong things, or love the right things in the wrong way. For example, they wrongly love power, fame, wealth, appearance, and many other things that are unworthy of our love. They also love things that are worthy of love, such as other individuals, but this is done in an excessive manner, putting others into a place reserved for God, this is a serious mistake and undermine happiness (Austin & Geivett, 2011).

Looking at the conception of happiness in Asian Culture we come across different philosophical thoughts of ancient China like Confucianism, Taoism, Buddhism etc. Confucianism originated about 600 years before the Christian era in the east of present-day China (Shangdong province). If someone has 'Jen', it is implied that he has 'virtue', and if someone has 'virtue', it is implied that he has reached the conditions required to live a good life. Followers of Confucianism must bring Jen into their own society, into their social relations. Every activity should be carried out in accordance with virtues such as benevolence, duty, property, intuition and trust, and every man can live up to the virtues and thus live a good life. According to Confucius, positive attitude towards learning is central to virtue, the more knowledge one has, the closer to virtue one gets. Learning should be a part of every place and every thing. Learning is the first necessary condition of a good life. Morality is a main requirement for a good life, not only for individuals but also for community as a whole. The humanities, arts, sciences and technology can also help us to reach a good life, but morality is more important (Zhang, Veenhoven, 2008).

Taoism, another important philosophy of ancient China, was also founded in the sixth century B.C. For Taoism, people should keep the balances between human beings and Nature, between individuals and society, between oneself and others. So the best way to have a good life is to follow the 'Tao', the 'Law' of nature. This will make people happy. According to Taoism, we cannot find the real conditions of a good life in the outside world; we can only derive them from our inner world, from our spiritual condition. A simple life is the best life. Taoism argues against the development of wisdom in people. For Taoism, children are the happiest and most healthy human beings. People are advised to follow the laws of nature and to avoid societal norms: people should wait for nothing, should desire nothing. When you give up all things, you have something left, that is happiness (Zhang, Veenhoven, 2008).

Feng shui comes from the Taoist ideals in order to deal with nature and it can be considered as one of the oldest approaches to human wellbeing. In principle feng shui searches for a harmony between human and the natural world. It is based on an emotional conception of nature, a system of superstition supposed to teach people where and when to build a tomb or erect a house so as to insure for those concerned everlasting prosperity and happiness. (Eitel, 1873) Chiou and Krishnamurti (1996) defines feng shui as the art of adapting buildings, rooms, and furniture so as to achieve a maximum of harmony with nature, particularly its local environment and climatic conditions. The Chinese believed that each place on earth has special topographic features that indicate the qi so called wind and water or Feng-shui. The place where breath accumulates is considered to be an ideal location for building a yang house (for living) or a yin house (grave). To keep this living breath, the form of the house must be able to store the wind. The concept not only determines the axis and location but also principles of form: symmetry, front and back, symmetrically located secondary buildings and so on (Chiou, Krishnamurti, 1996). In this perspective every location and every person is unique and the built environment needs to answer to these unique conditions with the consideration of laws of nature.

Another Asian philosophy, Buddhism originated in Northern India about six hundred years before the Christian era. Its founder was Buddha Gautama (560–480 B.C.). Buddhism focuses on suffering because only by understanding a problem can a solution be found. Buddhism recommends people to turn their attention to their personal, inner worlds, and to give up every illusory desire of happiness. Instead they should focus on the spiritual enlightenment. The Buddhist has to give up every thing, including one's own happiness (Zhang, Veenhoven, 2008).

Approaching the Middle Ages, the conceptions of happiness are characterized by important thinkers of the Islamic Golden Age. Greek philosophy, mostly of a late Hellenistic Aristotelian and neo-Platonic character, had been translated into Arabic and were being developed by a number of famous Muslim philosophers: al-Kindi (d. 252/866); al-Fanlbi(d.339/950); and especially Avicenna (Field, 1991). According to Avicenna (c. 980–1037), also known as "Ibni Sina" happiness is the aim of humans, and that real happiness is pure and free from worldly interest. In his different works, Avicenna has pointed to various types of euphoria, including the intellectual pleasures and the physical or sensual pleasures; he believes that the intellectual pleasures are well superior to the physical pleasures and says that a wise person should not limit himself to sensual pleasures, as animals do. (Raftari, 2015). For three reasons, the intellectual pleasure is superior to the sensual pleasure:

1. The sensual pleasure is changeable, alterable, and unstable, while the intellectual pleasure is durable and unchanging.
2. Intellectual pleasures result from the deep perception of things and phenomena, whereas sensual pleasures come from a superficial perception of appearances.
3. Sensual pleasures are limited and finite whereas intellectual pleasures are unlimited and infinite (Office of Islamic philosophy, 1999).

Different from the views of Aristo, Avicenna thinks that euphoria is something true and hence-belonging to the hereafter, which cannot be attained in this world. On the other hand similar to Aristotle, Avicenna thinks of euphoria in stages but the stages of Avicenna are different from those of Aristotle. For Aristotle, depending on the amount of his efforts during lifetime, man can increase his euphoria. For Avicenna Euphoria consists of different stages in a hierarchical order and the highest stage of man's euphoria is his similarity to the active mind: human souls are distinct from each other in terms of their access to knowledge or perfection or other virtues. The farthest stage of euphoria stems from the separation of spirit from the body. He states that the final euphoria of soul comes about upon the abstraction from the material world, and when the soul achieves the perfect abstraction (Raftari, 2015).

Ghazzali (c. 1058–1111) as another muslim philosopher contemporary of Avicenna saw the tendency towards Hellenistic philosophy as a positive danger to the Muslim community and attempted to discredit it in two major works, "The Intentions of the Philosophers" (Maqiisid al-falGsifa) and The Incoherence of Philosophy (Tahafut al-faliisifa). Ghazzali's influence also extended well beyond the Arabo-Persian Islamic world (Field, 1991). Parts of his work have since been translated into Turkish, Urdu, Pushto, Bengali, Malay, Hebrew, Latin, English, French, Italian, Spanish, German, and Russian, thus reaching a vastly larger Muslim and non-Muslim audience. The Revivification, written in Arabic, has naturally been Ghazzali's best known work in the Arab world and among Western scholars of Islam. His Alchemy is based on turning away from the world to God, and its constituents are four (Field, 1991): the knowledge of self; the knowledge of God; the knowledge of this world as it really is and the knowledge of the next world as it really is.

Mevlana can be seen as another muslim philosopher known all over the world for especially the Mevlevi Sama Ceremony (Whirling Dervishes) and his teachings that focus on tolerance and love. The key concepts in Rumi's teachings are as follows: certain acceptance and acknowledgement of both positive and negative experiences; unlearning of old habits and looking at the world with new eyes; decentering, changing one's focus from Self to Other; and attunement of body and mind through mediation, music and dance (Gretty, 2010). Although Rumi's teaching do not necessarily focus on happiness, it aims to maximize the inner potentials of individuals corresponding to the eudaimonic tradition that highlights positive psychological functioning and human development.

The Renaissance has an important role for the development of the individual in the society. Renaissance saw the rise of the individual as a powerful force in the world. This was result of the new form of culture, inspired by Greco-Roman literature, which they referred to with names like the studia humanitatis (the humanities), studia humaniora (more humane studies), studia honestarum artium (the study of honourable arts), bonae litterae (good letters), bonae artes (the good arts), erudition legitima et ingenua (noble and legitimate learning). This culture occupied a middle ground between purely practical studies such as law, medicine, or the mechanical arts on the one hand, and purely theoretical studies such as natural philosophy, advanced logical theory, metaphysics, and theology on the other (Hankins, 2007). Hankins discusses that the Humanists claimed that study of good letters made people better, more virtuous, wiser, and more eloquent. It made them worthy to exercise power and made them better citizens and subjects when not exercising power. Human studies embellished life, brought pleasure, and nourished piety. The humanities did not save souls, but living a good life would bring men favour in the eyes of God and strengthen piety, or at least not damage it.

In renaissance the individual was important but individuals rarely operated alone, however, groups of people commissioned architecture to serve their collective goals. With buildings like housing for elderly, orphans, schools and buildings for religious organisations groups could achieve communal goals (Anderson, 2013).

The Industrial revolution represents another breaking point offering significant changes in the role of the individual in the society. The effects on living conditions of the industrial revolution have been very controversial, and were hotly debated by economic and social historians from the 1950s to the 1980s. (Woodward,1981). Meanwhile, life expectancy increased from the late 18th century and stagnated from the 1810s onwards (Wrigley et al. 1997), that period also saw rising levels of inequality (Williamson 1985). According to Albarran (2016), in Britain (seen as the locomotive of the Industrial revolution), the well-being of an average British worker shows a steady upward trend throughout most of the period. From the beginning of the period until 1810, gains in health improve welfare, despite stagnant wages and longer working hours. In the subsequent years, life expectancy ceases to increase because of rapid urbanisation and its associated maladies. In that period, real wages become the main driver of welfare improvements. Welfare no longer increases after 1840, when productivity gains accelerated in Britain and their benefits trickled down to the working classes in the form of higher real wages. A closer look at the forces driving this process reveals that the health toll paid by a growing urban population together with long working hours and rising inequality hampered progress in well-being (Albarran, 2016).

Research on subjective measures of well-being can only be traced back to the 1920s and 1930s. According to Angner (2011), the appearance of a science of happiness was perhaps marked by Myerson's 1917 study promoting the discipline of eupathics, the more gracious sister of eugenics, focusing on the well being of the normal (Myerson, 1917). Systematic happiness research took off shortly thereafter. Subjective measures emerged in marital success studies, educational psychology, and personality psychology in the 1920s and 30s, as tools developed for the study of personality were applied to happiness and satisfaction; the measures were further shaped by the epidemiology of mental health, gerontology, and the social indicator movement in the 1960s and 70s, as measures of happiness and satisfaction were recruited as proxies for mental health and well-being in large, representative samples. The scientists invented a fascinating variety of measurement tools, some of which were ultimately rejected, but some of which turned out to have a remarkable longevity. A number of the studies were longitudinal; some used remarkably large, representative samples (Angner, 2011).

ARCHITECTURE FOR THE GOOD LIFE

All through history, architecture has been one of the most important tools of human to shape their living conditions and to communicate the values, which are more important for them. On the other hand, the priorities of architecture were different not always focusing on human needs and expectations.

According to Bishop (2015), if we are to describe a person with well-being, we would describe a host of objective and subjective facts about the person, including (1) positive feelings, moods, emotions (e.g., joy, contentment), (2) positive attitudes (e.g., optimism, hope, openness to new experiences), (3) positive traits (e.g., friendliness, curiosity, perseverance), and (4) successful interactions with the world (e.g., strong relationships, professional accomplishment, fulfilling hobbies or projects). So well-being is a result of positive way of thinking reflected to one's life and surroundings. In this sense well-being always exists with the environment we live in and it also has a dialectic relation with it. So, we can see the natural and built environment not only as an aspect that affects well-being but as a tool that promotes well-being. Therefore if we want to understand the conception of well-being in different times in history, architecture represents the most powerful clues letting us understand society's ways of living, their priorities and their ways of thinking.

Ideal Architecture for Well-Being in Theory of Architecture

Essays and guides over time do not only seek to answer questions about the origin and purpose of architecture but they also try to provide rules to combine theory and practice: a system capable of guiding the architectural composition. Among these, De architettura of Vitruvius (ca. 40 BC) is the oldest and then in the Renaissance Period, De re aedificatoria (1485 ed. completa) by Leon Battista Alberti and the Four books of the architecture of Andrea Palladio (ed. 1570) are equally important. Then come the treatises of the eighteenth century like Lodoli and Laugier but also Boullée with his 'Arquitectura. Ensayo sobre el arte' which had great influence on architecture, mainly in Italy. Afterwards we arrive first at the eclecticism of the nineteenth century and then at the dissolution of "an idea of architecture" unique in the twentieth century (dominated, in the first half, by opposing currents such as rational organic style expressionism). From the second half of the 20th century, different ideas of architecture coincide with the specific poetics of some prominent exponents: these are the cases of the"tendenza" by Aldo Rossi, the "postmodern" of Venturi and Scott and the Deconstructivism of Peter Eisenman.

Vitrivius says that without exaggerating, the architect should know something about everything enough to cope with different kinds of tasks, evaluate the relations and create the appropriate solutions. But in reality this also means that the answer that he has to give to life and being well is so complex that it affects every aspect. At this point, emerge the first concepts of well-being absolutely multifaceted and interdisciplinary. Vitrivius starts his ten books of architecture with giving his thanks to the emperor as he gives his attention not only to the welfare of society in general and to the establishment of public order, but also to providing public buildings intended for utilitarian purposes. With his words he intends that buildings are real complements of well-being especially in social terms and the understanding of well-being is not complete without the architectural scene. This also means that the politics of the time are also aware of the importance of architecture in the welfare of the society and they give the way to the creation of well-defined rules for ideal architecture.

Vitruvius' main contributions to the history of architectural theory include (1) his canonical account of the classical orders (Books III and IV), and (2) identification of three principles of architecture, *firmitas*, *utilitas*, *venustas*, conventionally translated as structural integrity, utility, and beauty; or (*per* Wotton 1624) firmness, commodity, and delight (Fisher 2016). For Vitrivius, utilititas has a preponderant role in architecture. Venustas is also important which means that in the relationship between two intervenes, the decorum must establish the ornamental forms appropriate to the subjects, the places and the circumstances. The concept of decorum is interesting: "refined aspect of the work", can be regulated by the statute, understood as "coherence with the state of a thing", a use therefore that derives not from customs, but from the very nature of that thing. Statute decorum is treating buildings with "coherence" with respect to their nature, i.e. to the dedication of a temple or function. Regarding customary decorum it is important to maintain coherence between choices. You cannot have a magnificent interior and a sloppy entrance or you cannot insert a decoration of a certain style into an element of another style. Natural decorum occurs when seeking consistency with natural aspects: orientation with respect to light, wholesomeness with respect to water etc. Also the concept of distribution is interesting, understood in different ways such as "economy" (for example making use of materials available on site and "accordance" (different buildings in the city according to their role, but also the different rooms inside a house, adapting it to each person (= matching of members to their use).

For Vitrivius the aspect of health and well-being in architecture starts to be considered with choosing the right site for the building. He discusses that the first thing to look for in places is their appropriateness for human health (which may not correspond to the needs of other animals). In this sense an architect must know about gnomonics in order to find the right orientation towards the sun. Moreover, medicine in Greek-Roman times had a field of investigation completely different from the modern one. Vitruvius in fact argues that an architect must have a basic notion of medicine "in view of the areas determined by land inclination, the properties of air and places, which may be healthy or unhealthy, and the waters". That is, medicine was a discipline that studied the climatic and physical properties of a place to determine which was the most suitable for the development of a new city.

Vitruvian influence on subsequent architectural theory cannot be overstated. Architectural theory as a didactic or polemic pursuit (beyond De Architectura, which Vitruvius may have intended as a documentary effort) was launched in the Renaissance with Leon Battista Alberti's De Re Aedificatoria (1485). Alberti offered a "recovery" of Vitruvius and an outward tribute to the Vitruvian triad as definitive pronouncement of architecture's aims, hence the measure of architectural success (Fisher,2016).

Leon Battista Alberti reinterprets the three Vitruvian categories (firmitas, utilitas and venustas) as: Firmitas: choice of land, materials to be used and foundations, Utilitas: various types of buildings in relation to their function and Venustas: architectural beauty as a harmony expressed mathematically thanks to the science of proportions. He tries to interpret the ancient theories in a modern way defining four essential points:

1. art is a science,
2. art must interpret an ideal goal of beauty,
3. art must deal with human actions,
4. art has the purpose not only to delight

Beauty involves the rational integration of the proportions of all parts of a building, so that each of them has a definite size and shape and nothing can be added or removed without destroying the harmony of togetherness. For Renaissance artists, harmonic perfection was therefore an absolute value independent from subjective perception. Alberti puts utilitas in the background compared to venustas: it means that a building that fulfils its function and is"comfortable" but not beautiful and elegant is worth very little. And in fact beauty contributes to the comfort and even to the duration of the building. So everything must be done to ensure that the work is not only functional and comfortable, but above all well adorned and pleasing to the eye. Compared to Vitruvius, therefore, Alberti gives greater importance to beauty, but his Venustas concept is a little bit different from that of Vitruvius: beauty is not only a private experience, which is in the relationship between building and user, but acquires a universal value: concinnitas, in fact, is "the fundamental and most exact law of nature". To clarify this concept Alberti re-evaluates Vitruvian principle of symmetry, and gives the metaphor of the building-body a much wider value: the harmony between the various parts of the human body is transferred to the architectural project and subsequently to the whole city, according to a progression from the intimate and private sphere of the house to the public sphere of the State.

Although the comparison between architecture and the organism, so important in the aesthetic of Alberti, derives from Vitruvius, we can observe fundamental differences between the two theories: L'Alberti welcomes the Vitruvian theory of the "commodulatio ratæ partis", that is the commensurability of every single member of the work and of all the members in their entirety together with a given unit of measure or module, but his conception of beauty goes beyond the "symmetria" of Vitruvius. In fact "concinnitas", even if having a value similar to that of "symmetria", refers to a relation that is the fruit of not only a sum of elements, but also of their reorganization as a whole', in this case assuming a meaning similar to "organicity". Therefore it recalls the aesthetic ideal of Cicero, of an organicistic and functionalist type, according to which the harmony found in the human figure is based on a condition of equilibrium. Thus, while the "symmetria" of Vitruvius can be modified to meet the needs of the beholder (eurythmia), the "concinnitas" of Alberti, on the example of Cicero, requires the immutability of the order, because it is based on a precise law as nothing can be added, removed or changed (Stefano, 2010). It is therefore clear that the concept of a building body analogy cannot be considered, as will be the case in all post-De re ædificatoria treaties, as a metaphor or a simple anthropomorphic projection. In Alberti, in fact, it assumes a broader semantic value. The constructions must take the model of the body of the living on the mainland, like the ships in sea and the shape of fish. In this analogy the theory of Vitruvius merges with that of Cicero, which defines perfection on the model of the human body where each part responds to a specific need, and in this perspective Alberti establishes the necessity of indissolubility of utilitas and venustas in architecture. Consequently abstract and metaphysical criteria as the geometric space also play an important role in the reflection of Alberti, corresponding to the principle that architectural form depends on the needs of man (Stefano, 2010). According to Edward R. De Zurko (1957), in the functionalist theory of beauty that is structured in analogy to the human body, where each element performs a function, there is a different beauty in each object depending on the purpose. But one of the most serious ambiguities of Alberti regarding the principles of beauty and functionality can be noted in not having properly distinguished the different types of purposes like individual pleasure, the well-being of the whole society or something intermediate.

The belief that architecture is science, and that each part of the building, both inside and outside, must be integrated into a single and identical system of mathematical relationships, can be called the fundamental axiom of Renaissance architects. The central problems of Renaissance architecture can be summarized in the symbolic meaning of architecture and proportional organization of buildings. The last among the great humanist architects who tackles these issues, in their theoretical and practical work, is Andrea Palladio. He publishes The Four Books of Architecture to discuss building materials and techniques; designs of private urban townhouses and country villas; city planning issues and designs of specific Roman temples. Wittkower in “Principi dell’architettura palladiana” (1964) argues that Palladio has always been consistent with the search for universal rules and precise norms to be used in design. Of course, the design model must be adapted to the particular needs of each task. Other authors, contrary to this interpretation, such as James Ackerman (2000), rejects the existence of an ideal archetype which exists a priori in Palladio’s mind, thus minimizing the importance of the geometric grid and proportions, and considers the idea of a principle that the skill of the architect transforms each time into a design of original villa. The Palladian idea is therefore to build also according to the surrounding context, reconciling visionary and ecological balance, building and maintaining a good quality of life around. In this sense he is the true interpreter of the

Renaissance, with the man at the centre of the circle of Leonardo, who is the arbiter of his own destiny.

At the heart of Palladian architectural conception lies the need for the right relationship faced with the use of harmonious relationships, not only within each environment, but also in the relationships of the environments among them: Palladio, following Alberti, accepts the mathematical definition of beauty. The fundamental novelty of Palladio's architecture is that it has developed an organic structure that results in systematically binding one room to another by means of harmonic proportions. These relationships, which other architects had used in correspondence of the two dimensions of a facade, or of the three dimensions of a single room, were employed by him to integrate an entire building.

Renaissance buildings were meant to be enjoyed at many levels. This period of intense intellectualism in the arts was also one of profound sensual delight. Aesthetic experiences through the mind as well as through the body were thought to increase the efficacy of a building whether its purpose was to display the power of the ruler or enhance religious practice. Further architecture was intended to fulfil the agenda of beauty, and beauty had a purpose; the choice of ornament, material and style could make a building more effective, better able to fulfil its function in the world (Anderson, 2013).

Approaching the eighteenth century, we come across the Lodoli's ideas in order to create a better architecture and thus overcome the past. In this sense Lodoli focuses on two aspects:

- Nothing shall show in a structure which does not have a definitive function, or which does not derive from the strictest necessity.
- There shall be no architecture, which does not conform to the very nature of the material.

When architecture will have attained these two great objectives, it will then be true, honest and reasonable. Lodoli set out to scrutinize architecture and lay bare the very bases of all building. In doing so, he came to feel that contemporary architecture was fundamentally wrong. This did not mean simply an attack on the Baroque but it repudiated as well all methods of building conforming to the principles enunciated by Vitrivius and reinstated during the Renaissance. Algarotti, coeval of Lodoli, grasped the great promise contained in Lodoli's thought but could not condemn what was to him supreme architectural beauty. Therefore he defended Baroque buildings but could not admit their perplexing contradictions. Certainly it was absurd to have a material denying its own nature, like stone carved like wood, but it was not easy to meet this charge of Lodoli. In this sense only the individual conscience of the particular artist could arrive at a decision. Man is the measure of things and what is made by man shall be made according to him and according to his will (Kaufman, 1944).

Laughier, best known for his Essay on Architecture published in 1753, discusses the natural language of architecture and begins with people in the state of nature. Among their needs is the need for shelter. Architecture in this view may be said to be both: the image of the cave and the image of the forest (Schulz, 1965; Harries, 1983). As Laugier presents this system, only columns, entablatures, and pediment representing the supporting uprights, the horizontal members they carry, and the inclined members that make up the roof are considered essential parts of architecture. Walls, windows, doors are permitted but are said to make no essential contribution to beauty. To Laugier, The Primitive Hut was the highest virtue that architecture should achieve and it represented the idea of architecture as the mediator between man and nature. Laugier believed that the model of the primitive man's hut provided the ideal principles for architecture in general and could be applied to any other building. Nature in this sense was the source for the general principles of architecture. The turn to the primitive hut does not mean a functional approach to architecture. What lifts architecture beyond mere building is its power of representation (Harries, 1983). According to Harries (1983), Laugier's "nature" speaks with a regional voice and his interpretation of this voice is very much shaped by historical situation. Therefore region and history help determine what is natural and hence inevitable.

In "Architecture, Essay on Arté, Boulleé criticizes all architecture in history. Architecture, for Boullée, is neither fantastic (Piranesi), nor is it arbitrary (Perrault). It is neither the art of building (Alberti), nor is it the science of construction (Vitruvius). Negating both the "functionalism" of Lodoli equally with the effete proportional debate of Perrault and Briseux, Boullée, grounds architectural harmony not in mathematical and abstract ratios of the orders, but rather in the simultaneous in situ apprehension of regularity, symmetry and variety, that is, in sensation (Dodds, 1995). For Boullée, regularity, which is a function of simplicity and clarity yields beauty while symmetry yields order. Variety, which, according to Boullée, is paradoxically "derived from the most perfect symmetry," yields a diversity of sight, "which, by its very nature, likes to be stimulated and sustained by what is new. He continues, "It therefore follows that variety puts new life into our faculties by offering us new pleasures..." The sphere is the proof of the paradox for Boullée, in that, the most infinite variety is derived from the most perfect symmetry. Giving examples from nature, from the human anatomy, he argues that symmetrical compositions are true and pure. The slightest disorder, the slightest confusion becomes intolerable. Order must be in evidence and paramount in any composition. In short, the wheel of reason should never desert an architect's genius for he should make a rule of the excellent maxim "Nothing is beautiful if all is not judicious". To give a building character is to make judicial use of every means of producing no other sensation than those related to the subject. Good taste is a delicate aesthetic discernment with regard to objects that arouse our pleasure. He defines great architecture with the following words:

As in nature, the art of giving an impression of grandeur in architecture lies in the disposition of the volumes that form the whole in such a way that there is a great deal of play among them, that their masses have a noble majestic movement and that they have the fullest possible development. The arrangement should be such that we can absorb at a glance the multiplicity of the separate elements that constitute the whole. The play of light on this arrangement of volumes should produce the most widespread, striking and varied effects that are all multiplied to the maximum. In a large ensemble, the secondary components must be skilfully combined to give the greatest possible opulence to the whole; and it is the auspicious distribution of the opulence that produces splendour and magnificence.

Boulle gives the Palace at St. Germain-en-Laye, the Metropolis and Newton's Cenotaph as examples through which he tried to convey with architecture the image of grandeur and he suggested the reader to consult the plans in place in order the get the true message as he believes that an artist is not to explain well but to execute well. The Newton's Cenotaph is noted as the building, which explains Boulle's ideas about the geometric style and talking architecture. The massive and spherical shape of the building was inspired by Boullée's own study called "theory of bodies" where he claims that the most beautiful and perfect natural body is the sphere, which is the most prominent element of the Newton Memorial (Perouse, 1974).

Arguing that good architecture is a function of good taste, Boullée shared Edmund Burke's belief that good taste had a fundamentally civic function, leading to the public good "sensus communis". The architectural distillation of this idea may be best exemplified by Boullée's project for a Stadium. Designed in accord with the principles of good taste, the stadium was intended to provide a civic function similar to that of the antique Roman coliseum.

Figure 1. Boulle, The Newton's Cenotaph

That is, the stadium was intended to be the locus of the citizenry's seeing itself within the decorum of a public gathering, within a tastefully produced architecture. Moreover, the goal of any public architecture, for Boullée, should be the construction of sensations, which are indistinguishable from those we experience in nature.

Boulleé concludes in Essays on Architecture by saying that a student of architecture needs to become a student of nature for he will be forced to recognize that this is the source of beauty in art. If there are ways of making man perfect, they can only derive from the study of nature.

The modern movement can be regarded as another important impact on the understanding of ideal architecture and the role of human in this context. The technological advancements offered numerous opportunities to architects in order to go beyond the limits of the past. Modernist architects considered historical styles as irrelevant and they tried to create a new style driven by both the opportunities and needs rejecting unnecessary detail and decoration. Sullivan known as the creator of the skyscraper, animated by certain understandings of self and community that would later define the Chicago School of Sociology, saw himself as a kind of social engineer whose architecture would generate democratic space and encourage authentic interaction (Abbott, 2000). Sullivan criticized Greek and Roman temples and Gothic cathedrals as designs that have no meaning in democratic society. Abbott discusses that Sullivan might have been too quick, however, in dismissing historical architecture as mystifications, produced by and for the powerful as the designs he found offensive cannot be divorced from the dynamics of power. These designs spoke to ultimate meanings, to the higher purposes according to which social interaction was oriented. Sullivan was proud of the tall building and saw it as the biggest opportunity for humankind in creating a democratic environment driven by the needs of man. As Sullivan put it in his famous essay "The Tall Office Building Artistically Considered":

The tall building is one of the most stupendous, one of the most magnificent opportunities that the Lord of Nature in His beneficence has ever offered to the proud spirit of man. We are on the highroad to a natural and satisfying art, an architecture that will soon become a fine art in the true, best sense of the word, an art that will live because it will be of the people, for the people, and by the people.

A virtual consensus exists today in characterizing Sullivan's "proud and soaring" towers as antithetical to democratic principles and aspirations (Habermas, 1985) (Figure 2). The skyscraper illustrated man's control over the environment. According to Abbott, to confront this problem is to confront questions about the good or goods. To ponder what a democratic space might look like and what kinds of interaction this space might create

Figure 2. Berenice Abbott. Manhattan Skyline I, South Street and Jones Lane, 1936, 20,3 × 25,4 cm, The New York Public Library.

demands first an articulation of ideals, of higher purposes and ends towards which democratic selves aspire. He discusses that individuals were no longer organically related to their physical environment and were thus incapable of communing with it. They were no longer related to each other as the soaring towers atomized those who lived and labored among them. So from this perspective, the opportunities offered by modernism carried architecture to a point detached from the opportunities of nature and social interaction.

FUTURE RESEARCH DIRECTIONS

Future research needs to focus on the reflections of the conception of the good life on architecture and interiors at different times so that we can evaluate the relationship between well-being and space and compare it with today's understanding of well-being related to the environment.

CONCLUSION

Living a good life has been important all through history but ways of achieving it depended on society's ways of living and architecture was a follower of coeval ways of thinking. In this sense theory of architecture at different times tells us that the concept of the good life and happiness changed, evolved sometimes as a response to previous conceptions, sometimes as a response to societal and technological changes and sometimes as an evaluation of the offerings of nature.

Evaluating conceptions of happiness and the good life, it can be assumed that in time the role of the individual and its independence gained more importance while in ancient times the individual's happiness was regarded as a phenomenon more related to the society. The appropriate role of every individual in ancient societies was the key to his own happiness and to the happiness of others with the principle of "one for all, all for one". With the rise of Renaissance the individual gained more importance and it was seen as a powerful force in the world. This was result of the new form of culture, inspired by Greco-Roman literature. Then the Industrial Revolution represented another breaking point with the technological advancements and numerous opportunities offered to architects in order to go beyond the limits of the past resulting with a distancing of the individual from his social and physical environment.

Regarding theory of architecture and the search for ideal architecture that supports the concept of the good life, the first written principles were expressed by Vitruvius with his canonical account of the classical orders and identification of three principles of architecture, firmitas, utilitas, venustas. In Renaissance, Alberti offered a "recovery" of Vitruvius giving more importance to beauty (Venustas) defining it as a rational integration of the proportions of all parts of a building, so that each of them has a definite size and shape and nothing can be added or removed without destroying the harmony of togetherness. Then Palladio as a true interpreter of the Renaissance, gives importance to build also according to the surrounding context, reconciling visionary and ecological balance, building and maintaining a good quality of life around. Approaching the eighteenth century Lodoli set out to scrutinize architecture underlining the importance of functionality and accordance with the nature of material. Negating the "functionalism" of Lodoli, Boullée grounds architectural harmony not in mathematical and abstract ratios of the orders, but rather in the simultaneous in situ apprehension of regularity, symmetry and variety. With the rise of modernism, the tall building expressed the opportunities of technology. The creator of the skyscraper Sullivan saw this new architecture as a generator of democratic space that would encourage authentic interaction. On the other hand with the continuous rise of individualism in time, people were detached from each other and from the natural environment.

Today we know that well-being is highly subjective and although living in similar conditions, individuals reported levels of well-being can be very different. As it is impossible to evaluate the well-being of individuals from different times we can only evaluate their writings and the clues related to their living environments, which we find in architecture as a concrete and conscious construct. In this sense the evaluation of architectural structures is the best source from which we can read the history of the concept of the good life. The search for the good life finds its concrete reflections on the search for good architecture.

REFERENCES

Abbott, J. R. (2000, Spring). Louis Sullivan, architectural modernism, and the creation of democratic space. *The American Sociologist.*

Ackerman, J. S. (2000). *Palladio. Piccola Biblioteca.* Einaudi.

Albarrán, D. G. (2016). *British well-being 1780-1850: Measuring the impact of industrialisation on wages, health, inequality, and working time*. CEPR Policy Portal. Retrieved from https://voxeu.org

Anderson, C. (2013). *Renaissance architecture*. Oxford University Press.

Angner, E. (2011). The evolution of eupathics: The historical roots of subjective measures of well-being. *International Journal of Wellbeing*, *1*(1), 4–41. doi:10.5502/ijw.v1i1.14

Austin, M. V., & Geivett, R. D. (2011). *Being good: Christian virtues for everyday life*. Eerdmans Publishing.

Bishop, M. (2015). *The good life: Unifying the philosophy and psychology of well-being*. Oxford University Press. doi:10.1093/acprof:oso/9780199923113.001.0001

Boullée, E. L. (1953). *Architecture, essay on art* (H. Rosenau, Ed.). Tiranti.

Bradburn, N. (1969). *The structure of psychological well-being*. Aldine.

Chiou, S. C., & Krishnamurti, R. (1997). Unraveling Feng-shui. *Environment and Planning. B, Planning & Design*, *24*(4), 549–572. doi:10.1068/b240549

Diener, E. (1984). Subjective well-being. *Psychological Bulletin*, *95*(3), 542–575. doi:10.1037/0033-2909.95.3.542 PMID:6399758

Dodds, G. (1995). *On Boullée's nature: Scattered beauty and the seeing image*. Retrieved from www.academia.edu

Eitel, E. J. (1873). *Feng-Shui: The rudiments of natural science in China*. Trübner & Co.

Field, C. (1991). *The alchemy of happiness: Abu Hamid Muhammad al-Ghazzali (Revised and Annotated by Elton L. Daniel)*. Routledge.

Fisher, S. (2016). Philosophy of architecture. In The Stanford Encyclopedia of Philosophy. Metaphysics Research Lab, Stanford University.

Habermas, J. (1988). Modern and Postmodern Architecture. In J. Forester (Ed.), *Critical Theory and Public Life*. MIT Press.

Hankins, J. (2007). *Humanism, scholasticism, and Renaissance philosophy*. Cambridge University Press. doi:10.1017/CCOL052184648X.003

Harries, K. (1983). Thoughts on a non-arbitrary architecture. Perspecta 20: The Yale Architectural Journal, 9-20.

Jackson, N. J. (2013). Exploring subjective well-being and relationships to lifewide education, learning and personal development. Lifewide Learning, Education & Personal Development e-book.

Kahneman, D., Diener, E., & Schwarz, N. (Eds.). (1999). *Well-being: The foundations of hedonic psychology*. Russell Sage Foundation.

Kamtekar, R. (2018). Marcus Aurelius. In The Stanford Encyclopedia of Philosophy. Metaphysics Research Lab, Stanford University.

Kashdan,T.B.,Biswas-Diener,R.,&King,L.A.(2008).Reconsideringhappiness: The costs of distinguishing between hedonics and eudaimonia. *The Journal of Positive Psychology*, *3*(4), 219–233. doi:10.1080/17439760802303044

Kaufmann, E. (1944). At an Eighteenth Century crossroads: Algarotti vs. Lodoli. *Journal of the American Society of Architectural Historians*, *4*(2), 23–29. doi:10.2307/901186

Lyubomirsky, S., & Lepper, H. S. (1999). A measure of subjective happiness: Preliminary reliability and construct validation. *Social Indicators Research*, *46*(2), 137–155. doi:10.1023/A:1006824100041

Mirdal, G. (2010). Mevlana Jalal-ad-Dın Rumi and mindfulness. *Journal of Religion and Health*, *51*(4), 1202–1215. doi:10.100710943-010-9430-z

Myerson, A. (1917). Eupathics: A program for mental hygiene. *The Journal of Abnormal Psychology*, *12*(5), 343–347. doi:10.1037/h0073646

Office of Islamic Philosophy. (1999). *Avicenna's epistle concerning love*. Tehran: Author.

Pappas, N. (1995). *Plato and the Republic*. Routledge.

Pérouse de Montclos, J. M. (1974). *Etienne-Louis Boullée: 1728-1799: Theoretician of revolutionary architecture*. Thames and Hudson.

Raftari, H. (2015). Happiness in view of Aristotle and Avicenna. *International Journal of Social Science and Humanity*, *5*(8), 714–719. doi:10.7763/IJSSH.2015.V5.545

Rogers, C. R. (1961). *On becoming a person*. Houghton Mifflin.

Ryan, R. M., & Deci, E. L. (2001). On happiness and human potentials: A review of research on hedonic and eudaimonic well being. *Annual Review of Psychology*, *52*(1), 141–166. doi:10.1146/annurev.psych.52.1.141 PMID:11148302

Ryff, C. D. (1989). Happiness is everything, or is it? Explorations on the meaning of psychological well-being. *Journal of Personality and Social Psychology*, *57*(6), 1069–1081. doi:10.1037/0022-3514.57.6.1069

Simeonov, T. (2011). *Plato's Ideal City-State*. http://scott.london/articles/idealcity.

Stefano, E. D. (2010). Il "De equo animante" di L.B. Alberti: una teoria della bellezza. In F. Furlan & G. Venturi (Eds.), *Proceedings of Leon Battista Alberti, Gli Este e l'Alberti: Tempo e misura* (pp. 15–26). Fabrizio Serra Editore.

Waterman, A. S. (1993). Two conceptions of happiness: Contrasts of personal expressiveness (eudaimonia) and hedonic enjoyment. *Journal of Personality and Social Psychology*, *64*(4), 678–691. doi:10.1037/0022-3514.64.4.678

Williamson, J. G. (1985). *Did British capitalism breed inequality?* Allen & Unwin.

Wittkower, R. (1964). *Principi architettonici nell'eta' dell'umanesimo*. Einaudi.

Woodward, D. (1981). Wage rates and living standards in pre-industrial England. *Past & Present*, *91*(1), 28–46. doi:10.1093/past/91.1.28

Wrigley, E. A., Davies, R. S., Oeppen, J. E., & Schofield, R. S. (1997). *English Population History from Family Reconstitution 1580-1837*. Cambridge University Press. doi:10.1017/CBO9780511660344

Zhang, G., & Veenhoven, R. (2008). Ancient Chinese philosophical advice: Can it help us find happiness today? *Journal of Happiness Studies*, *9*(3), 425–443. doi:10.100710902-006-9037-y

Zurko, E. R. D. (1957). Alberti's theory of form and function. *The Art Bulletin*, *39*(2), 142–145. doi:10.1080/00043079.1957.11408374

Chapter 3
Well-Being in Practice and Policy

ABSTRACT

Recent research focuses on the concept of well-being, aiming to systematize it and obtain design guidelines. In latest years, various building certification systems have arisen, which, although used for ex post evaluations, contain, specularly, design guidelines. In a first phase the concept of well-being was intended on a global scale, linked to the pitfalls of pollution and consumption of resources, so design guidelines and control systems developed within the construction industry to ensure the conservation of the environment and therefore the "well-being" and "health" of human communities. Having therefore developed certification systems measuring and evaluating the performance of buildings in relation to their impact on the environment and its resources, we are now faced with a shift of attention on a smaller scale, linked to the performance that buildings offer not so much with respect to the environment as to the people who live in them. This chapter explores the concepts behind such systems and the relationship between building certification systems and people's well-being.

DOI: 10.4018/978-1-7998-4231-6.ch003

INTRODUCTION

Well-being has progressively been used as an important goal in many contexts, conditioning strategies, policies, and disciplinary approaches. In the architectural field, at different scales, this concept is crucial because the physical environment can strongly affect the conditions of life.

If we consider health, according to the World Health Organization (WHO), as "not merely the absence of disease or infirmity", but as "a state of complete physical, mental and social well-being" (WHO 1948), we can argue that people are not healthy in absence of spiritual dimension, related to a broader perspective including several conditions.

Although research on 'well-being' has grown in the last two decades, assuming a leading role in government policies around the world (e.g. Diener et al 1999; Kahneman et al 1999; Keyes et al 2002; Stratham e Chase, 2010; Seligman, 2002, 2011), it has not been accompanied by appropriate research and theoretical formulations of well-being, and still the issue is not resolved, because there have been many confusing and excessively broad definitions of well-being (Forgeard et al 2011).

Furthermore, the concept of well-being has been used in almost an interchangeable way with Quality of life (QoL) and comfort, as important goals in many different contexts. Although several theories have been developed concerning each concept, the lack of clarity about differences exists (Sara Pinto et al., 2017).

THE CONCEPTS OF QUALITY OF LIFE, COMFORT, WELL-BEING

The concept of Quality of life (QoL) has been discussed in several disciplines over time. The majority of authors define the concept as the individual's perception of their personal situation in their own life in the physical, social, mental and spiritual dimensions (Sara Pinto et al., 2016). The first studies emerged in the 1930s, and significantly grew in the post-war period.

The World Health Organization Quality of Life Group defines it as the "individuals' perception of their position in life in the context of the culture and value systems in which they live and in relation to their goals, expectations, standards and concerns". In this definition a person's all conditions are considered, including beliefs, environment and culture: QoL is therefore a very subjective concept, and it has been particularly studied in the nursing field. In Peplau's theory, (Forchuk 1993), QoL is a synonym for well-being or psychological health, but Leininger (1970) believes that the concept of QoL is culturally constructed.

The term "comfort", in a linguistic dictionary, is a synonym of "well-being" (Oxford dictionaries 2016) and it has always been a central concern and concept in architecture.

Referring to a general condition of health, Kolcaba defines comfort as "the immediate state of being strengthened by having the needs for relief, ease, and transcendence addressed in the four contexts of holistic human experience: physical, psychospiritual, sociocultural, and environmental" (Kolcaba 2003). Ida Jean Orlando (Marriner-Tomey and Alligood 2006) defines comfort as a response to human needs, while Sister Callista Roy (Roy 1976) stressed the importance of psychological comfort and Hildegard Peplau stated that comfort is a fundamental human need associated with food, rest, sleep and communication.

Regarding behavioural psychology, and in particular Alasdair A.K. White's Theory of Comfort Zone (White 2008), comfort indicates the mental condition in which the person acts in a state of absence of anxiety, with a constant level of performance and without perceiving a sense of risk. It is a condition of routine, familiarity and safety that makes us feel at ease without taking risks or perceiving any danger. In order to understand how complex the concept of comfort is, and how much it can even hinder well-being, the theory of the "comfort zone" shows how often individuals want to preserve their comfort (as defined above) to such an extent that they persevere in conditions that do not fully satisfy their desires and ideals, precisely because they are afraid of losing that state of quiet. With reference to psychology, therefore, the tendency to preserve one's own mental comfort could turn out to be like a real prison: it can be opposed by forcing oneself to carry out new and different activities, to change one's usual behaviour and so on.

However, it should be remembered that the term "comfort", in Architecture, holds two fundamental meanings: one intended as sensory comfort (therefore related to temperature, light, sound and so on), the other related to the ergonomic characteristics of the objects and furniture that people use (which can be more or less "comfortable", i.e. make everyday life easier and "cozier"). It can be defined as a condition of satisfaction in which an individual stands and does not wish to leave: it is actually a mental condition of satisfaction related to the environment.

Also the term "well-being" entails a multidimensional concept, usually related to happiness, positive experiences or ideas, life satisfaction, pleasure and prosperity.

Ereaut and Whiting (2008) argue that well-being is a cultural construct, because it is nothing more than what people agree is a good life (cf. Jackson 2013).

Then we can argue that the three above mentioned concepts are partially superimposable, and are definitively complex and multifaceted, but we can also identify some specificity: as Pinto and colleagues illustrate, distinct attributes can be observed for each concept: comfort seems to be more related to the reduction of imbalances or discomfort, inner peace, security and effective communication, while the concept of well-being lies on a psycho-spiritual basis, associated with happiness and an "internal energy" and QoL appears as a broader concept, associated with life improvement, dignity and achieving autonomy and personal goals. (Pinto et al. 2016). Both QoL and comfort are also defined as the state of satisfaction across physical, psychological, social and spiritual dimensions (although it may depend on the individual's perception, values, beliefs and culture).

Pinto and colleagues recognize that the concept of well-being can be assimilated to comfort, but they underline that the etymological origin of the concept of comfort leads to an action or intervention (to comfort), while the etymology of well-being refers to a state (of the body and of the spirit). Furthermore, they found that comfort tends to be used in a more holistic perspective while well-being tends to be more confined to the psychological or spiritual dimensions.

In other words, well-being tends to be presented on a psychological, social and spiritual perspective rather than on a physical perspective, like comfort is. Despite the holistic nature of the concept of comfort, the environmental perspective seems to represent its unique characteristic.

In conclusion, the Pinto and colleagues study highlights the paradoxical nature of the concept of well-being: on one hand, the use of the concept essentially concerns the psychological and spiritual dimensions, and, on the other hand, the concept is used in an almost indistinct way in the concepts of comfort and QoL ((Pinto et al 2017).). These three concepts are also related to the health definition proposed by the WHO, and are raised as an attempt to measure the subjective dimensions of health.

At the end of this short overview, for the aforementioned concepts three different nuances are here proposed from a purely architectural perspective and not from a medical or nursing one.

Comfort is therefore intended as a situation more or less lasting in time but in any case temporary, relative to a precise situation in which, thanks to design interventions of various nature, the user can carry out his or her activity without disturbing or unpleasant elements, from a physical and perceptive point of view. In this sense, comfort is closely connected with architecture, intended as environment, spaces and furnishings.

The quality of life is intended as an average among the different components of an individual's life as usually it is: first of all the (work or pleasure) activities that he/she carries out, and then the degree of satisfaction that they provide him or her, but also how much stress he or she usually has to face, how much his or her life is consistent with his or her inclinations, and of course how much the environments in which he or she lives and works are comfortable and aesthetically pleasing. Furthermore, it is a subjective concept, as the personal satisfaction is its most important component. Architecture therefore influences QoL to the extent that it provides comfortable spaces, facilities and furnishings, but it certainly cannot, on its own, guarantee a good quality of life as a whole.

Well-being is here intended as more similar to QoL, but it is expressed with reference to a specific and temporary situation, that could evolve in relatively short time. It is then an inner and subjective state, affected by comfort and QoL level, but referred to a certain time, to a certain situation, to a certain physical environment or to a certain activity. Despite a high level of comfort and QoL, people could the same feel the absence of well-being, and vice versa. The well-being is felt subjectively, and depends on culture and age, but it has to meet also several "objectives" requirements. Like QoL, it's clear that also well-being cannot be ensured by Architectural spaces only.

Since, in general, the concept of well- being refers to working spaces and is applied in order to increase people's productivity, assuring them comfortable environments (Singh & Syal 2010, Xue et al 2016), it should be pointed out that in this context well-being will instead be mainly addressed to the living spaces, a topic that has been considered secondary in this field.

WELL-BEING FRAMEWORK

While considering major factors that contribute to well-being, it is important to differentiate those that can be controlled: space can be regarded as the most important concrete aspect that can be controlled by the designer and can directly affect human well-being. Regarding the environment as a whole, a focus on the interiors (where human spend a very high percentage of their time and perform a great part of their basic activities) is needed.

Figure 1. Well-being Framework for Interiors
(Saglar Onay, Minucciani, 2018)

As stated above, well-being is extremely subjective and it is very difficult to measure with objective criteria. On the other hand, designers need to handle the issue of well-being with a methodological approach, as it is very difficult to achieve well-being with an intuitive approach (onay, Minucciani 2018). There are several requirements, strongly connected to space and context, that need to be fulfilled in order to promote user's well-being.

In design practice, indicating specific guidelines aimed at well-being might seem a tautology, since in effect the entire design process should be aimed precisely at this, at the comfort of inhabitants and users. Nevertheless, at the beginning of the process the designer needs to define the design criteria including contextual and human related criteria, and during the design process he/she needs to choose a design approach to handle the design criteria with a holistic understanding.

A methodological framework (fig. 1) can help the designer during these two vital steps. It summarizes all the requirements related to different components of well-being, with specific reference to interior architecture.

The proposed framework therefore identifies different areas in which well-being is perceived: from a purely physical and sensorial point of view, from a social and psychological point of view, from a functional and ergonomic point of view and from an aesthetic point of view. All these requirements are in relation to each other and they are subject to cultural differences regarding users with different backgrounds. Then the cultural context is determinant not only for the perception of well-being but also for its very meaning.

WELL-BEING AND SUSTAINABILITY

Based on previous discussion, well-being intended in a broad and holistic sense is certainly an objective that architecture pursues and has always pursued: precisely the evolution of the concept of well-being and the awareness of its complexity accompanies the evolution of architecture itself, at all scales.

Certainly it can be stated that, in a wider and more comprehensive notion of sustainability, the concept of well-being in living spaces, where people spend most of their time, should be a fundamental factor and a real "contributor to sustainability" (Saglar Onay & Minucciani 2018). However, in recent times the focus on sustainability has centered on a "collective" well-being, even beyond the present generation, then on the well-being that only the preservation and respect for the environment can ensure.

This has led to a focus on buildings and their performances (especially in terms of energy and chemical performance), being concerned about reducing their negative impact on the natural environment and thus ensuring sustainable living conditions for present and future generations.

At the same time, the advent of the Humanities and their disciplinary contaminations with sciences such as architecture induced us to consider the effects that urban decisions have on communities and architectural choices have on individuals. Although it may seem obvious that architecture aims at the people's well-being, in fact, guidelines specifically aimed at this objective in the light of recent theoretical developments in sociology, psychology and physiology have not yet been defined (Potrˇc Orecht et al. 2019). Then the practice has pursued this objective according to two sets of resources: the first concerns creativity and sensitivity of the designer, from which the idea of the project derives; the second concerns the large set of rules and regulations that the project must comply with.

Some of these standards are purely technical and structural in nature, but most of them have a more or less marked influence on the well-being of those who will benefit from architecture. Today, a scientific and deterministic approach aims to control and manage even a concept as broad, complex and (above all) subjective as well-being. Therefore, neither the concept of well-being has been excluded from the "certification season" that has opened a few years ago, particularly referred to sustainable architecture and outlined below.

In practice, if designers were used to starting with an aesthetic and functional idea, and verifying it complied with the regulations - now certifications try to attract the project's attention on elements and objectives that are politically decided (because they bring a social and economic advantage). The key to ensuring designers' attention is an evaluation process at the end of which the building may or may not be certified. This result is not a purely formal accreditation, but allows access to economic benefits and advantages, as well as ensures a higher value of the building within the real estate market.

However, the original concept that characterizes the architectural project must not be lost, sacrificing it to respect the certification requirements: while currently it may be possible that it makes it lose.

The relationship between building and sustainability is much more complex than simply meeting requirements that guarantee respect for the natural environment and the saving of resources: speaking of "Architecture" - and not simply "Construction" - attention to the physical, social and psychological well-being of individuals is also an element that conditions the sustainability of the project. Petermans and Pohlmeyer discussed the potential of Interior Architecture to contribute to subjective wellbeing and pointed to the value of applying a human-centered perspective in the design process (Petermans and Pohlmeyer 2014), introducing the Positive Design framework (Desmet and Pohlmeyer 2013).

In conclusion, it is necessary to balance the attention to sustainability between external environment and internal spaces, between the well-being of future generations and the well-being of current citizens: "It is particularly important to emphasize the fact that the sustainability of a building is reflected in its full occupancy and that such a level of occupancy can only be achieved by ensuring the health and well-being of the occupants. (Potrˇc Orecht T. et al. 2019).

CERTIFICATION SYSTEMS

The interest in the field of health and well-being of users of buildings increased in the last decades. As mentioned above, in order to define parameters to be respected and to evaluate the effects of design on well-being (although the concept may have different meanings as we have seen), in recent decades systems for the evaluation and certification of buildings have been developed.

Since these systems were oriented towards the assessment of buildings for the purpose of granting a certificate (and not to provide design guidelines), a holistic assessment could only be given in a second phase, after having disarticulated the different components of well-being, which was on the other hand largely intended as physical health.

Technological innovations played an important role in these certification systems, but they were still evaluated according to the optimization of objective parameters of comfort especially in relation to sustainability (intended as energy saving and in general related to the performances of a building).

The main difficulty lied in the fact that "well-being" was not intended as a continuous flow of inner states and activities connected to each other, related to the environment and to the context, but precisely as the sum of single parameters in a very complex whole. Therefore, the problem of the interaction between the different components and, probably, of the corrective factors (that can hardly be composed in a complex algorithm) remains.

In the preparation of a project, a certification system more than an "a posteriori" certification should be a body of guidelines able to influence the initial phase of the concept, as well as the aesthetic and functional idea of the designer. If this is not the case, one would risk aiming at a sort of standardization to which the uniqueness of the architectural artifact should bend.

Many certifications try to force the attention, in the different phases of the project, on elements and objectives that are politically decided. For this reason, the objectives of the certification must be clear, but it must be very flexible in its components, in order not to be a forcing: what counts is in fact the global result, intended from the point of view of well-being.

In any case, today they aim to build sustainable buildings from different perspectives: social and economic ones included. As Potr˘c Orecht and colleagues acknowledge, while the evaluation of economic and environmental aspects for buildings has become common, the social aspects have only recently gained importance all over the world: to label a building as sustainable, the health of its occupants has to be evaluated (Potr˘c Orecht T. et al. 2019).

If the primary objective of evaluations is the environmental sustainability, it tends to take on a global dimension, and even when it is intended at local level, it still remains a concept of territorial extent, changing in geographical space but also in historical time. If, on the other hand, the question of well-being is addressed, it may also be intended as a collective quality of life but the individual dimension, for which architecture works, cannot be ignored. This is why a "good design" could be inadequate, as the objectives are variable with the geographical and cultural context, and the most important thing is the final result. Since hardly all requirements can be completely fulfilled, it is important to understand that the goal is not their complete achievement but rather the pursuit of a special, architectural identity that preserves the original holistic concept.

In the system of certifications, in fact, the quality of interior spaces intended in their architectural and spatial connotations is poorly considered, then these systems risk mortifying architecture in its truest essence. Interior architecture is too often considered as a practical discipline, instead of a technical one but (before being technical) it must be cultural and theoretical: a discipline of research, practice and education.

So we need to be aware that certification systems have appropriated the concept of well-being by making it seemingly richer (i.e. by attributing a huge amount of parameters to it) but in truth it could become poorer: for this reason, research is even more important.

Therefore, a first generation of certifications focused on the performance of the building, as mentioned above. They were linked to a concept of sustainability closely intended as respect for the environment and resources: this is the case of the most popular one, LEED®, a voluntary certification program that classifies the energy efficiency and "ecological footprint" of buildings throughout their life cycle. The LEED (Leadership in Energy and Environmental Design) system was created and developed by the U.S. Green Building Council (USGBC) and the first version was launched with the help of the Federal Energy Management Program in August of 1998. It evaluates the performance of the building from an energy, water and emission point of view and takes into account the ecology of the materials and their location in relation to the environment. Therefore, the entire building, and not its single components, will be subject to certification, based on performance parameters. Providing a set of measurement standards for assessing environmentally sustainable construction LEED is now applied in more than 40 countries in addition to the United States. It is an open and transparent process where the technical criteria proposed by the LEED committees are publicly reviewed for approval by more than 10,000 organizations that are part of the USGBC.

LEED system focuses on perhaps the most dramatic problem related to the construction sector (i.e. the consumption of land and resources, as well as pollution), it promotes a sustainability oriented approach, recognizing the performance of buildings in key areas, such as energy and water saving, reducing CO2 emissions, improving the ecological quality of the interior, the materials and resources used, the design and choice of site. The system is based on the attribution of 'credits' for each requirement. Like similar others, it only partly covers the issue of health and safety of occupants and neighbours, as well as accessibility (Griffiths 2018). The elements to take into account are mainly quantitative, defining the final score, while qualitative parameters are assessed with the on/off criterion.

Actually, the LEED criteria were created to define the concept of "green building" by establishing a common measurement standard, to give recognition to the leaders of the construction industry who pay attention to the environment, to stimulate competition in the development of green projects, materials and construction methods, to raise awareness of the benefits of "green construction", as well as transform the construction industry (achieving the highest possible profit while maintaining the global ecology aspect of the project) and the real estate market (as the LEED certification gives buildings significant added value, it has quickly become an important parameter used by the real estate market).

It cannot really be stated that LEED is a certification system aimed at well-being, but at the most at health, and although it includes some parameters concerning the quality of interior spaces, it basically refers to the external environment and to the natural context, intended as a system of limited resources and more or less favorable to the health of living beings.

Like the majority of BCSs (Building Certification Schemes), LEED uses credit criteria among which energy remains the most important one, but also the Indoor Environment Quality (IEQ) and water are considered crucial (Illankoon et al 2017). In particular, IEQ could be related to health and well-being of inhabitants, but few parameters really deal with it.

In conclusion, nothing in the LEED certification system is expressly aimed at the holistic (and subjective) concept of well-being, and no guidelines are designed to do so. Indeed, LEED can be considered as a system that attempts to combine the well-being/health of individuals (given by high quality housing) and the well-being/health of wider communities, where the more or less virtuous behavior of individuals have a potentially devastating or propitious impact.

While the first generation of sustainability has concerned the construction parameters of buildings, energy saving and the impact on the environment, the new frontier of sustainability aims precisely to personal "well-being", i.e. the health and wellness of people, beyond the mere technical and structural features of buildings: in other words, a certification system arises switching from performances offered by the building in connection with the environment, to performances offered in connection with the inhabitants (i.e., persons).

However, the introduction of certification systems often generates a reversal of roles: instead of focusing on the positive or negative effects that certain innovations or practices may have on well-being, the process is forced to introduce practices or innovations that are considered positive for well-being.

Despite the apparent tautology, the difference is meaningful. The greatest difficulty lies in understanding well-being as a continuous flow of related activities, which generates the problem of interaction between the many components of the project.

In European context, the leading sustainability BCSs are considered BREEAM, LEED, DGNB and HQE, that is mainly focused on the French market (Potrˇc Orecht T. et al. 2019). Usually the BCSs access parameters dependent on the management of the building as well as parameters dependent on the building design.

More radical is the Living Building Challenge protocol that comes from Seattle and was created in 2006 by Amanda Sturgeon to certify a niche of buildings, built "like trees": autonomous, 100% natural, capable of regeneration as nature does. The revolution is that this certification represents, at the same time, a real philosophy for a new generation of houses. As Langner points out, "although equity has historically been included within the LBC's requirements, awareness of its importance has been growing" and new frontiers of certifications and standards have to address also inclusion, beauty, and biophilia as equal requirements to energy and water standards (Langner 2019). LBC consists in seven "petals", each of which takes into account one aspect of how the house is developed: from the place to the management of resources, the structure must prove to be completely self-sufficient and totally integrated into the surrounding area. The first cases validated in Europe are the visitor center of a park in the UK, a residential structure in Spain and a single-family house in Italy. To measure and evaluate the quality of life a building offers, WELL (Well Building Standard) Certification has been created. It addresses what WHO calls SBS, sick building syndrome, that is the place where people live and work in conditions of stress, disease, tension: it follows and goes beyond the LEED Certification, which aims to ensure the quality of construction for the entire life cycle, from the design phase up to implementation. WELL, lunched in 2014 and managed by the WELL Building Institute (IWBI) focuses on the well-being of users, thus shifting the focus on the people who live in the building: this is a novelty for the European scenario. Environmental sustainability, the protection of the landscape, the use of renewable resources for energy production are certainly not aspects to forget or set aside (Rick Fedrizzi, president and CEO of IWBI, is also responsible for the dissemination of the LEED Certification protocol around the world), but they should be complemented by a type of approach, which is defined as holistic, that reconsiders people as the main driver of any type of activity, working or not.

Other systems emphasized these topics as crucial, to ensure the health and well-being of the occupants: Living Building Challenge, created by the International Living Future Institute in 2006; Health, Well-being and Productivity in Offices, by the World Green Building Council, and FitWell, by the Centre for Active Design. They aim to integrate the existing BCSs and among them, WELL is the most well-known certification system.

Potrˇc Orecht and colleagues analysed "which topics related to health and well-being of the occupants are represented in specific BCSs that deal with the sustainability of already constructed buildings… and to what extent they are processed", comparing the schemes with BCS WELL, that assesses already-constructed buildings (Potrˇc Orecht et al 2019). They studied LEED O+M (Building Operations and Maintenance), BREEAM In-Use und DGNB, based on a market study for the European Commission. In conclusion, they state that the common sustainability BCSs "in general lack a comprehensive and thorough approach" and the aspects of health and well-being. While some issues, already well known are addressed to a certain extent in all of them, the topics (related i.e. to fitness, and state of mind) are less represented or are not represented.

Several parameters are found in both LEED and WELL, in fact they evaluate how the building satisfies the well-being of people but without neglecting the use of renewable resources, environmental sustainability and landscape protection: actually, the understanding of well-being is slightly different.

PRINCIPLES OF WELL CERTIFICATION

The concepts underlying in the WELL Certification, which starts from the assumption that the time spent in closed environments is about 90% of a day and therefore should be lived in the best possible conditions, can be applied to all type of buildings (residential, school, hospital) but above all they are aimed at work environments not only because people live there for many hours every day, but also because a condition of serenity and well-being can have positive repercussions on the productivity of the entire Company.

The WELL Certification holistic approach is based on scientific evidence and performance analysis resulting from six years of research work. The WELL protocol version 1 aims to evaluate and therefore certify seven basic criteria (identified and developed by a team at Columbia University consisting not only of architects and engineers, but also of doctors, psychologists, nutritionists and anthropologists): air, water, food, lighting, movement, comfort, psychophysical well-being. These areas are further divided into topics, and the innovation component is also important for everyone.

Each parameter is analyzed through scientific measurements, so that the stakeholders can return clear data, in order to plan any strategies for improvement.

Actually, some of these seven elements can be measured with objective and therefore "irrefutable" parameters, while others cannot. In fact, if the ambition to apply a holistic approach rightly leads to consider psychological comfort, it is very difficult to say that it can be measured according to both objective and valid for everyone criteria.

In fact, WELL Certification evaluates the well-being not only considering the body, but also the mind and the spirit. Therefore, the psychophysical well-being is defined as a condition of peace and serenity, combined with the reduction of the possibility of getting sick. It is fostered by a greater link between nature and the human being.

The "movement" component in the WELL Certification has a role that previously had not been considered enough in indoor environments. The sedentary nature of daily life is notoriously linked to health problems, while constant movement guarantees a condition of psychic well-being and can be stimulated by appropriate design measures.

Similar considerations can be made about light, which not only has an objective functional value but also, often, a very strong influence on health, serenity and energy. Strategies are therefore encouraged to maximize natural light, reducing artificial light to the amount actually needed for the activity that takes place in the different rooms of the building. The same can be said about the air: its quality and constant exchange have a significant impact on well-being, not only psychological but also physical. The control of ventilation, humidity and temperature is therefore essential, as well as the absence of toxic materials such as constant maintenance of the systems dedicated to the exchange of air and filters.

Less related to architectural design, but similarly important and considered in the WELL evaluation, are the aspects related to a healthy and balanced diet, to the quality of water and its availability.

An important peculiarity of the WELL protocol is that it provides also the figure of a "well assessor" or reviewer, with the task of performance verifications on site – and each property is evaluated throughout its entire life cycle, from design to testing, every year and for three years even after its subsequent and actual occupation.

In addition to a mere effect on physical health, the benefits of such an approach are also aimed at a good mental state and at a greater link with nature (in all its forms), but the emphasis - in most cases - is on increasing work productivity (fewer illnesses, greater confidence in one's superiors; reduction of absenteeism at work; increase in teamwork; passion for one's own work...).

Allowing to measure and improve the well-being of people living and working in a building, and increasing their productivity, this novel protocol matches the principles of the so-called "organizational well-being" understood as the ability of the company to maintain the highest levels the physical, psychological and social health of workers. Achieving this Certification is a "proof of leadership", with a parallel increase in social responsibility perceived in the Company and in the market.

Once again, therefore, particular emphasis is placed on the economic advantages, which consist - in addition to the significant advantage for employers - also in a better real estate valuation of buildings WELL certified, which can increase their value by about 10% compared to buildings without environmental certification (from a survey carried out by Rebuild with CBRE and GBCI Europe, Rebuild 2020).

THE CONCEPT OF COMFORT AND WELL-BEING IN WELL PROTOCOL

One of the areas of greatest interest in relation to interior architecture is certainly comfort. WELL v1 considers thermal, acoustic, ergonomic, and olfactory comfort, clearly stating the intent of optimizing indoor environments, with particular reference to working spaces (WELL, 2020). Design for ergonomic safety, preventing physical stress, certainly improve comfort that is here defined through a series of parameters ranging from noise to cleanliness, from temperature to functionality of the furniture, from colors to materials: all these components have an influence on the physical and mental health of people, as well as their quietness, and in a way that is really perceptible.

Actually, as we've seen above, the concept of comfort is largely masterable by architectural design, but it is only one component of well-being. For its part, the WELL protocol identifies comfort with the psychophysical well-being of people living in a space (home, office, school) and with a feeling that depends on certain environmental conditions that can be partly planned and that consequently fall under the responsibility of the designer during different phases of the project (from concept to the management of the project itself).

However, in WELL's structure there are some subtle field overlaps that make the concepts of comfort and well-being not perfectly distinct. In fact, if it is stated that comfort also has positive consequences on the psychological and mental component, it must be recognized that each individual has a personal perception, and establishing certain parameters and thresholds is difficult. Furthermore, well-being is highly subjective, and - with reference to the well-being framework – it cannot be considered apart from the cultural context.

Nevertheless, WELL aims to encompass and overcome these difficulties. Biological responses to light and sound conditions, as well as to colours and materials, have been carefully considered so that, for example, the "light" component is not considered from the point of view of pure functional illumination, but is related to circadian cycles and to the quality of light (artificial rather than natural).

The Fitness concept in WELL is related to the risks of physical inactivity, and the role of buildings and urban environment is significant: not only outdoor spaces, but also indoor ones can encourage movement and physical activity. A specific area is dedicated in WELL to the "mind". It involves moods and mental states fostered by environmental and spatial conditions, that can impact stress, the quality of sleep, and so on. Elements such as the adaptability of spaces, the inclusion of nature, the sense of security strongly affect the mind state. WELL then gives conducting surveys about the satisfaction of inhabitants.

Undoubtedly, the mind area is particularly complex. According to WELL holistic approach underlined above, the state of mental health is closely linked to contrasts of order/disorder, light/dark, cleaning/dirt, hot/cold temperature, bright and disturbing colors/soft and relaxing colors. Evaluating the quality of the interior space, studying its level of comfort, WELL defines whether that space is noisy, clean, thermally and acoustically insulated, whether the furnishings are inserted and chosen in such a way as to be functional, whether the colors used are stimulating for the user and so on. The mental state and the human mood can change in time according to situations, and it also depends on physical health. In this sense, building design can strongly support it, offering psychological and physical benefits for occupants (Chan and Liu 2018).

As it can be seen clearly, WELL protocol combines in the same category parameters that were instead considered under different ones in the well-being framework presented above. Thermal comfort, for example, is a sensory requirement while the order/disorder condition refers to a requirement that, depending on the case, may be aesthetic or functional.

In spite of the declaredly holistic approach, however, WELL works by parcellarizing the concept of well-being and by defining comfort specific attributes: thermal comfort, acoustic comfort, light comfort, distribution comfort and so on. As a result, WELL protocol evaluates all these aspects but above all aims to improve and solve, where present, problems through the adoption of materials that guarantee an excellent performance from the point of view of both thermal and acoustic insulation, a correct plan layout and a functional partition of areas, an adequate choice of colors and furnishings and a valid lighting system that ensures the right amount of light both during the day and at night.

Actually there are at least two considerations to underline. The first concerns concepts of comfort: if limits of environmental values (of any kind) can be established with a certain precision, without causing a situation of discomfort, i.e. it is relatively easy to indicate what comfort is NOT as the positive definition of the parameters for comfort is not very clear and precise.

Although less complex than well-being, comfort is also a holistic concept. In fact, comfort cannot simply be identified as the sum of several components to be satisfied, and it can even be said that a situation of considerable and global overall comfort can be achieved despite one or more objective comfort parameters not being respected. The same can be said for well-being as a whole taking into account that it is also a variable condition depending on the cultural context, and we can understand how evaluation and certification systems can present some very difficult points of interpretation.

FUTURE RESEARCH DIRECTIONS

As stated above, well-being is a holistic concept, made of complex interactions among several parameters whom scientific measurement not always is possible and internationally sharable. Actually, involving also psychological needs and individual lifestyles, well-being is differently perceived in different contexts. The typical approach of Interior Architecture entails a strict relation with clients leading to a real customization, then it should be more considered in developing international Certifications: they are instead targeted to a global audience, tending to be deterministic while in this field the subjective component is strong.

Certainly, the WELL approach could also appear too deterministic in a field in which the subjective element is undeniable: a specific section of Certification systems should be studied to consider the role of the cultural context, that cannot be dealt with using standard values or principles.

Finally, also following the very complex openings that architecture has in this period with humanities, the emotional quality of space, which always exists, has to be remembered. It has to be particularly underlined: space is always emotional, and it is never "neutral".

The emotional "atmosphere" of interior spaces, which in recent years has been brought to the fore by famous architects and scholars (including Peter Zumthor and Juhani Pallasmaa), plays a very important role on the mood of the occupants, and has very close correlations with the functional destinations of the interiors. Actually WELL doesn't ignore this component: "Because the mind plays a vital role in an individual's overall health and well-being, an atmosphere that supports a healthy mental state can have significant psychological and physical benefits" (IWBI 2019). However, this topic is poorly addressed.

Under the "Mind" category many features are listed. Among them, the "Beauty and Design" that focuses on aesthetics qualities of meaningful space has a positive impact on occupant mood and mitigating stress. WELL is featured by an extensive richness of details, including a large number of parameters and elements. With respect to the aesthetic qualities of spaces, for instance, general principles are given on the heights of the ceilings, the proportions of the corridors, shapes and colors, integration of artworks. Another component very important in this topic is the relationship with nature: incorporation of natural element as well as interaction with them to improve the mind status.

While the concepts are actually useful and complete, suggested solutions could create some conflict or interference with the design process. However, it should be pointed out that the scope of WELL is related to existing buildings and that the assessment is therefore given a posteriori. Therefore, it is not a question of project guidelines: however, the demarcation line could become very thin over time.

CONCLUSION

The International WELL Protocol and other Certification Systems testify the contemporary attention to individual well-being as a sustainable goal, overcoming the mere physical parameters. In the same time, the definition of parameters and values according to a very disarticulated system might conflict with the holistic nature of well-being concept.

The role of the cultural context must be taken into account not only in the assessment of a building and its spaces, but also in defining basic concepts such as comfort or quality of life: they're concepts that have many overlaps with well-being, but aren't perfectly synonymous. With reference to the proposed framework, sensory and ergonomic requirements may be scarcely influenced by the cultural context, but all the others may be subject to significant variations: in particular social and psychological requirements depend on cultural context, that is they depend on the habits and roles that are assigned to citizens in a given society, but also expectations and even the notion of well-being (as well as quality of life) also highly depend on them. The same can be said for the aesthetic parameters, which are rooted in different cultures and their history, as well as the connection to natural context e the relationship with natural elements derive from centuries of established habits and lifestyles.

Moreover, as the WELL Protocol repeatedly points out, there are unconscious and physiological reactions, which could be defined as chemical and biological, to the stimuli offered by spaces and atmospheres. Neuroscience tells us that the first reaction to a space is not conscious but emotional, and that it influences behavior in that space for a certain time. Therefore, when you enter a new atmosphere, it is the body that first feels the stimuli coming from its surroundings and, after a temporary disorientation, you "perceive" how you feel, reaching a new balance (Böhme 2001).

For this reason, after having experienced several generations of anthropometry (the purely physical and static one, the dynamic and ergonomic one, the cognitive one), Architecture now probably has to face, with the help of neurosciences, the "emotional anthropometry" - although it may seem an oxymoron.

Only the designer's sensitivity can master and manage all these aspects, which no normative framework can ever assure. Certification systems are certainly useful as a compendium of all issues, requirements, and interactions between project components - and as assessments of existing buildings for possible improvement works - but they could not (and should not) become guidelines for the design process.

REFERENCES

Alligood, M. T. (2006). *Nursing Theorists and their work*. Mosby Elsevier.

Barcaccia, B., Esposito, G., Matarese, M., Bertolaso, M., Elvira, M., & De Marinis, M. G. (2013). Defining quality of life: A wild-goose chase. *Europe's Journal of Psychology*, *9*(1), 185–203. doi:10.5964/ejop.v9i1.484

Böhme, G. (2001). *Aisthetik: Vorlesungen über Ästhetik als allgemeine Wahrnehmungslehre*. Wilhelm Fink Verlag.

Callista, R. (1976). Introducrion to nursing. An adaptation model. Prentice-Hall.

Chan, I. Y. S., & Liu, A. M. M. E. (2018). Effects of neighborhood building density, height, greenspace, and cleanliness on indoor environment and health of building occupants. *Building and Environment*, *145*, 213–222. doi:10.1016/j.buildenv.2018.06.028 PMID:32287986

Desmet, P. M. A., & Pohlmeyer, A. E. (2013). Positive design: An introduction to design for subjective well-being. *International Journal of Design*, *7*(3), 5–19.

Fitwel. (2020, April 20). Retrieved from https://www.fitwel.org/

Forchuk, C. (1993). *Hildegarde E. Peplau: Interpersonal nursing theory – Notes on nursing theories (10)*. Sage.

Griffiths, K., Boyle, C., & Henning, T. F. P. (2018). Beyond the Certification Badge—How Infrastructure Sustainability Rating Tools Impact on Individual, Organizational, and Industry Practice. *Sustainability*, *10*(4), 1038. doi:10.3390u10041038

Illankoon, I. M. C. S., Tam, V. W. Y., Le, K. N., & Shen, L. (2017). Key credit criteria among international green building rating tools. *Journal of Cleaner Production*, *164*, 209–220. doi:10.1016/j.jclepro.2017.06.206

International Living Future Institute. (2020). *Living Builing Challenge*. Retrieved from https://living-future.org/lbc

IWBI International WELL Building institute. (2019). *The WELL Building Standard*. Retrieved from https://www.wellcertified.com/

Kolcaba, K. (2003). *Comfort theory and practice: a vision for holistic health care and research*. Springer Publishing Company.

Langner, E. (2019, May 10). *The Living Future unConference Emphasizes Equity in Sustainable Design*. Retrieved from https://www.metropolismag.com/architecture/living-future-unconference-2019-review/

Leininger, M. (1970). *Nursing and Anthropology: Two Worlds to Blend*. Wiley.

Lyubomirsky, S., Sheldon, K. M., & Schkade, D. (2005). Pursuing happiness: The architecture of sustainable change. *Review of General Psychology*, *9*(2), 111–131. doi:10.1037/1089-2680.9.2.111

Naci, H., John, P. A., & Ioannidis, M. D. (2015). Evaluation of wellness determinants and interventions by citizen scientists. *Journal of the American Medical Association*, *314*(2), 121–122. doi:10.1001/jama.2015.6160 PMID:26068643

Obrecht, T. P., Kuni, R., Jordan, S., & Dovjak, M. (2019). Comparison of health and well-being aspects in building certification schemes. *Sustainability*, *11*(9), 2616. doi:10.3390u11092616

Petermans, A., & Pohlmeyer, A. E. (2014). Design for subjective well-being in interior architecture. *Proceedings of the 6th Annual Architectural Research Symposium*, 206–218.

Pinto, S., Fumincelli, L., Mazzo, A., Caldeira, S., & Martins, J. C. (2017). Comfort, well-being and quality of life: Discussion of the differences and similarities among the concepts. *Porto Biomed. Journal*, *2*(1), 6–12. doi:10.1016/j.pbj.2016.11.003 PMID:32258577

Rebuild Milano. (2018, October 29). *Il valore dell'innovazione*. Retrieved from https://www.rebuilditalia.it/en/MS/green-building-valori-e-tendenze/

Saglar Onay, N., & Minucciani, V. (2018). Wellbeing Framework as a Contributor to Sustainability. In *WIT Transactions on Ecology and the Environment, Volume 2* (Vol. 17, pp. 699–710). WIT Press.

Singh, A., Syal, M., Grady, S. C., & Korkmaz, S. (2010). Effects of Green Buildings on Employee Health and Productivity. *American Journal of Public Health*, *100*(9), 1665–1669. doi:10.2105/AJPH.2009.180687 PMID:20634460

WELL. (2020, April 20). *Well Certification*. Retrieved from https://www.wellcertified.com/

Well-Being. (n.d.). In *Oxford dictionaries*. Retrieved from http://www.oxforddictionaries.com/

White, A. (2008). *From comfort zone to performance management*. White & MacLean Publishing.

World Green Building Council. (2016, October 25). *Building the Business Case: Health, Wellbeing and Productivity in Green Offices*. Retrieved from https://www.worldgbc.org/news-media/building-business-case-health-wellbeing-and-productivity-green-offices

World Health Organization. (1948). *WHO definition of health*. Retrieved from https://www.who.int/about/definition/en/print

Xue, F., Gou, Z., & Lau, S. (2016). Human factors in green office building design: The impact of workplace green features on health perceptions in high-rise high-density Asian cities. *Sustainability*, *8*(11), 1095. doi:10.3390u8111095

Chapter 4
Different Design Approaches for Well-Being

ABSTRACT

Interior spaces have a great effect on human well-being because of their role as the follower of life. Flexibility becomes an important design issue while considering the subjective character of well-being in interiors. Therefore, not only the inhabitants' needs change during their lifetime, but especially interiors with multiple or changing users need to be considered with a design approach that can lead to multiple solutions that can be adapted to changing users and preferences. This chapter aims to evaluate a series of design approaches that are particularly attentive to a holistic conception of human well-being and that can be adopted with particular reference to interior design. Here it is not intended to identify a general hierarchy of such approaches, as their appropriateness depends on the context, on the user and on the interpretation of the designer himself. Therefore, it is important to know the potentials of different approaches in solving design problems with different priorities.

INTRODUCTION

Although architectural space can play a significant role on people's well-being, studies on the relation between space and well-being are very limited. More than the built environment the positive effect of the natural environment has been considered to be very important during the last decades.

DOI: 10.4018/978-1-7998-4231-6.ch004

Well-being in interiors has started to be discussed recently in many researches, articles and books. On the other hand we can say that all these works can be regarded as initial attempts to understand the relationship between well-being and space. Even the concept of well-being itself is defined in very different ways according to different perspectives.

Well-being can be defined as the positive state of mind and body in relation to the complex system of interconnected components that built up living. In literature conceptions of well-being appears to follow two main traditions which are the approaches following hedonic and eudaimonic traditions. If we look at the concept of well-being from the perspective of the hedonic tradition, design can promote pleasurable experiences and block undesired experiences. So every design object that involves us in desired activities can contribute to our well-being. If well-being is understood as actualizing one's human potentials so that he can make the appropriate contribution to the society, the role of design can be defined as creating the right tools with which a person can develop himself and find the right ways to perform his actions for his specific contribution. Architecture and spatial design has the potential to make considerable contributions to both hedonic and eudaimonic conceptions of well-being. While the hedonic tradition is concerned with more aesthetic and sensory aspects of space, the eudaimonic tradition necessitate the consideration of functional, social, psychological aspects related to space. So, looking from the perspective of design, the framework of well-being needs to consider all these different dimensions of well-being as the hedonic and eudaimonic views complete each other in terms spatial needs.

Within the field of behavior-influencing design for wellbeing, two approaches currently stand out that explicitly aim to change people's way of living and to improve wellbeing (Dorrestin & Verbeek, 2013). A first approach is the Persuasive Technology approach, as developed by BJ Fogg (2003). Persuasive technologies aim to guide the behavior of users in desirable directions, and are often concerned with issues in wellbeing, like health, hygiene, and environmental issues. A second contemporary approach to behavior-influencing design for wellbeing was developed by Richard Thaler and Cass Sunstein. In their book Nudge (Thaler & Sunstein, 2008), the authors make a case for designing our material surroundings in such a way that they influence us in a positive sense, without taking control away from us. The central idea in their approach is that the choices people make are to a considerable extent organized and pre-structured by our material environment (Dorrestin & Verbeek, 2013).

While we consider wellbeing in interiors, the most important contribution can be about structuring an understanding of design that methodologically supports the concept. This can be possible by evaluating the interior design process in a way that the design criteria handles all the issues that are necessary for wellbeing in interiors. Like every design activity, interior design also requires certain design input to be evaluated during the design process. As interior design is concerned with creating spaces that support human activities and well-being, the design input is closely related to the human factor as well as potentials of the existing natural and built environment. Therefore at the beginning of the process the designer needs to define the design criteria including contextual and human related criteria. Then during the design process, he needs to choose a design approach to handle the design criteria with a holistic understanding. As the outcome of the process, interior environments become generators of human wellbeing (Figure 1). In this sense, the issue of wellbeing in interiors needs to be examined especially in terms of design

Figure 1. Interior design process leading to wellbeing

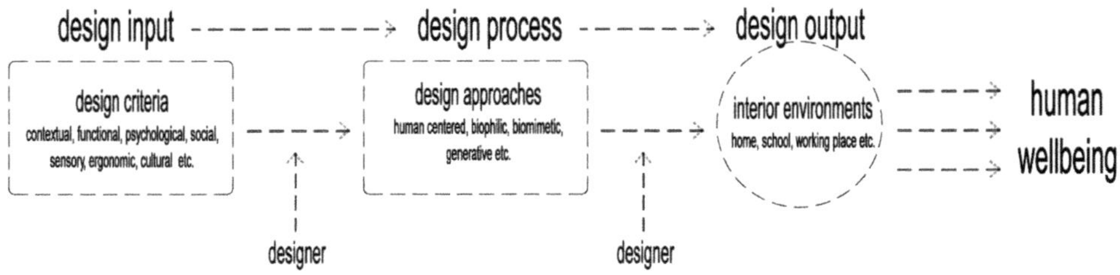

criteria and design process proposing a methodological framework that can help the designer during these two vital steps.

The design criteria in interior design needs to be structured considering the functional, psychological, social, sensory, ergonomic and contextual requirements of the users defined in the wellbeing framework for interiors (Saglar Onay, Minucciani, 2018) (Figure 2). The details of the wellbeing framework for interiors are discussed in Chapter 5. It is always important to note that all these requirements are subject to considerable changes related to the cultural background of users. Therefore developing spaces that support wellbeing is strongly related to understanding the users, the society, the

Figure 2. Wellbeing framework for interiors
(Saglar Onay, Minucciani, 2018).

specific environment and the living circumstances together with evaluations of the past and projections for the future.

After understanding user's needs and defining the requirements, the designer needs to get involved in the design process which is generally a multi-criteria design task because of the complexity and subjectivity of wellbeing needs. In this sense, it becomes extremely important to determine the design approach that will guide the designer to the design output. There are already different design approaches that have been developed to contribute to the quality of life of users and inhabitants. Although their main focus is not particularly on wellbeing, they have the potential to contribute to the concept of wellbeing related to space. Therefore the existing approaches in the field of architecture and design will be discussed in terms of their design input, design process and design output relationships and their potentials to contribute to interior design. Among these, human centered design is an approach widely used for design with its focus on creating things deeply based on general natural characteristics and peculiarities of human psychology and perception. Biomimicry and biophilic design are approaches mostly used in architectural design and they consider the relationship with nature as a factor that leads to increased well-being. Other approaches like supportive design focus on well-being in healthcare institutions so that the environment can contribute to the healing process of patients. Approaches like inclusive design consider the differences in user needs focusing also on groups with disadvantages and disabilities. The following section will focus on different design approaches and their potentials in leading to well-being related to interior space.

DESIGN APPROACHES FOR WELL-BEING IN INTERIORS

A design approach is about the way the design input is evaluated and processed. During the design process, the designer needs to decide on his priorities and create a certain hierarchy between the design criteria. Depending on the elements that the design approach considers crucial, field research focuses on elements that may also differ significantly from one case to another, with different priorities. In this part of the study, different approaches will be discussed from a theoretical point of view, considering their design process. Although the number of approaches to be examined may be increased, here the selection has been derived from their common use in literature to meet well-being requirements, with the exception of generative design. On the other hand generative design is more deeply discussed because of its potential to be used for well-being studies (Minucciani, Saglar Onay, 2018).

Biomimicry

For a long period, nature has been outclassed by human technology, but after the Second World War a specific attention has been paid to the role of nature in the design process. Relation between nature and human well-being could be intuitive, but the design approaches that can be derived from nature vary considerably according to different interpretations of different characteristics of nature.

In the last decades the nature was again looked at not only as a possible source of inspiration but also as a guide. This is just the case of biomimicry, term originated from the work of scientist Otto Schmitt, that literally means the imitation of life. According to Kennedy's definition (Kennedy 2004), biomimicry "refers to studying nature's most successful developments and then imitating these designs and processes to solve human problems", while Michael Pawlyn defines biomimicry as "mimicking the functional basis of biological forms, processes and systems to produce sustainable solutions"(Pawlyn 2011), Julian Vincent as 'the abstraction of good design from nature" (Vincent 1990) and Janine Benyus as 'the conscious emulation of nature's genius' (Benyus 1997). Benyus argues that a full emulation of nature engages at least three levels of mimicry: form, process, and ecosystem (Benyus, 2008).

Biomimicry is the examination of nature models, systems, processes and adaptations as answers to specific functions, mechanical and structural problems - and consequently their adoption in architecture, generating individual products ("organs") or systems and processes ("organisms") (Minucciani, Saglar Onay, 2018). Thus it usually means the creative implementation of biological concepts into design process. Nevertheless, often it is simply intended as direct copying of nature, and actually some scientists preferred the term 'biomimetics'. But biomimetics in this sense cannot be juxtaposed, from an architectural perspective, to 'biomorphism' that is mainly interested to biological forms by replicating them, while biomimicry is specifically focused on developing sustainable solutions, and its final solution may or may not look organic, or visually resemble the organism from which the lesson came. Obviously, it cannot automatically produce good architecture, especially if it encompasses a purely scientific approach to design. Human well-being calls into question also the emotional, spiritual, even historical dimension (Minucciani, Saglar Onay, 2018).

In interior architecture, the common, formalistic use of biology as a library of shapes has to be overcome by understanding the rules governing the natural forms instead of simply replicating them. Another term related to this trend is 'bio-utilisation', that refers to the direct use of nature to achieve positive effects on buildings (e.g. incorporating plants to gain evaporative cooling) (Minucciani, Saglar Onay, 2018).

Biomimicry design, although strictly connected to our quality of life and even to our survival, is not specifically well-being-oriented, as its main goal is the sustainability through the emulation of life's engineering. Sustainability is devoted to a long term vision, where well-being is the result of peaceful coexistence with the natural world (Minucciani, Saglar Onay, 2019).

Looking to the organism, architecture applies its form and functions to a building. An almost obvious reference is the Norman Foster's Gherkin Tower (2003) with a hexagonal skin inspired by a sponge, as well as the Grimshaw Architects' Eden Project (2001) in Cornwall, that consists of domes modelled after soap bubbles and pollen grains.

On the other hand, building can mimic how the organism interacts with its environment: the Eastgate Centre designed by architect Mick Pearce in Harare, Zimbabwe, has no air-conditioning or heating but a passive cooling system, inspired by the self-cooling mounds of African termites, regulates its temperature (although the structure doesn't appear as a termite mound).

Another example is the Qatar Cacti Building, designed by Bangkok-based Aesthetics Architects for the Minister of Municipal Affairs and Agriculture: it is inspired by the cactus's relationship to its environment. Finally, biomimetic design can also be inspired by an ecosystem, mimicking how the different components work together in the same environment.

Evaluating the successful examples in architectural design, biomimicry can be used as a problem solving methodology to create new sustainable standards for interior environments. This may include the overall design of interiors as well as interior components or various interior systems such as heating or lighting. For example bioluminescence is a more readily available approach in biometric design; there are several designers and companies working to develop viable means of illumination by microorganisms. There are also growing number of bio-inspired materials and products expanding the friendly environment options available to designers. With the advancements in computational design, computer aided manufacturing and digital sensor technologies; an innovative materially oriented design approach becomes possible, which enables the design and manufacturing of surface structures responsive to climate without need for any mechanical or electronic control (El-Zeiny, 2012).

Nevertheless, biomimicry approach could be adopted to solve well-being needs: its design process starts from the observation of natural world - seeking out a local organism or ecosystem that fulfills the function- then "talking" with it, asking and listening it to emulate its design and process (Figure 3). If we realize that nature is able to respond to the basic needs of creatures, not

Figure 3. Biomimetic design process

only simply solving problems, then we can also acknowledge that well-being needs can be fulfilled by drawing inspiration from nature and above all by observing it: in fact, this approach implies a careful capacity of observation and deep understanding of the vital processes (Minucciani, Saglar Onay, 2018).

Biomimicry can contribute significantly to wellbeing in interiors if it is handled as the general approach in the whole interior design process, but not only materials or secondary systems. This necessitates a deeper analysis of the problem to be solved and identification of similar problems and the ways they are solved in nature.

Biophilic Design

Following E.O. Wilson definition of Biophilia as "the innate tendency to focus on life and likelife processes" still under evaluated (Wilson 1984), R. Kellert states that this human inclination "to relate with life and natural process is the expression of a biological need", encompassing an ethic attitude toward the world and being biologically based. He suggests that human identity and personal fulfillment somehow depend on our relationship to nature." (Kellert- Wilson 1993).

This human need for nature affects our emotional and cognitive development: consequently, Biophilia is strictly related to human performance and well-being. Even if some scholars consider biohpilia as a branch of biomimicry, they are quite different approaches. While biomimicry focus on practical solutions inspired by natural forms, processes and systems, biophilia describes humans' connection with nature and biophilic design is replicating experiences of nature in design to reinforce that connection. Biomimicry goal is in achieving better performance, while biophilic design aims to improve health and well-being. Biomimicry is more heavily used in technology and product development, while biophilia concerns more directly to interior design, as well as to architecture and urban design. Then these two concepts address nature in different ways, biomimicry recognizing the innovation potential of natural solutions, biophilia recognizing the health and well-being benefits of connectedness with nature (Minucciani, Saglar Onay, 2018).

Biophilic Design aims to create healthy and productive habitats supporting well-being as a whole: "Biophilic design is the deliberate attempt to translate an understanding of the inherent human affinity to affiliate with natural systems and processes—known as biophilia —into the design of the built environment" (Kellert- Wilson 1993). Thus, it is an innovative way of designing places going beyond green architecture that focuses on decreasing the environmental impact of the buildings, but this is insufficient to benefit to well-being nor to reconnect us to the natural world (Minucciani, Saglar Onay, 2018).

A growing body of data and knowledge supports the role of contact with nature in human well-being. R. Kellert (2005) notes that contact with nature has been found to enhance recovery from illness, that people living in proximity to open spaces report fewer health and social problems as well as superior quality of life and a stronger sense of place, that office settings with natural environmental features improve worker performance and motivation (Kellert 2005). These studies provide scientific support for the assumption that contact with nature is critical to human well-being (Minucciani, Saglar Onay, 2018).

According to R. Kellert, our species is biologically related to natural world based on nine different valuations of nature: utilitarian, naturalistic, ecologistic-scientific, aesthetic, symbolic, humanistic, moralistic, dominionistic, negativistic. These cover very different approaches, from emotional attachment to alienation – encompassing very different functions, from mental development to peace and to security: but, although in different ways, our relationship with nature remains therefore unavoidable.

There are three overarching health responses in biophilia that help explain how individuals interact with their environment: cognitive, physiological and psychological. More recently, Heerwagen (2006) laid out a framework for "features and attributes of buildings linked to well being needs and experiences" reflecting these relationships in human-centric terms; and according to Cramer and Browning (2008), human-nature relationships tend to fall into three broad experience categories: nature in the space, natural analogues, or nature of the space. Seven patterns related to 'nature in the space' interactions have been identified as: visual connection with nature, non-visual connection with nature, non-rhythmic sensory stimuli, access to thermal and airflow variability, presence of water, dynamic and diffuse light, and connection with natural systems. Natural analogues are objects, materials, colors, shapes, patterns and algorithms that evoke nature. Broadly speaking, analogues can be characterized in architecture and design as representational artwork, ornamentation, biomorphic forms and natural materials. Three 'natural analogue patterns' have been identified as: biomorphic forms and patterns, material connection with nature and complexity and order. 'Nature of the space' refers to different spatial configurations and associated psychological and physiological responses they engender. Four 'nature of the space' patterns have been identified as: prospect, refuge, mystery and risk/ peril (Ryan, Browning, Clancy, Andrews & Kallianpurkar, 2014).

Kellert (2008), defines Biophilia with two basic dimensions and six elements related to 70 biophilic design attributes. The first basic dimension of biophilic design is an organic or naturalistic dimension, defined as shapes and forms in the built environment that directly, indirectly, or symbolically reflect the inherent human affinity for nature. The second basic dimension of biophilic design is a place-based or vernacular dimension, defined as buildings and landscapes that connect to the culture and ecology of a locality or geographic area. This dimension includes what has been called a sense or, better, spirit of place. The two basic dimensions of biophilic design can be related to six biophilic design elements: environmental features, natural shapes and forms, natural patterns and processes, light and space, place-based relationships and evolved human-nature relationships. These six elements are tl1en revealed in more than 70 biophilic design attributes. For example some attributes related to environmental features are air, water, sunlight, colour, natural materials, plants and animals. And some of those in relation to natural shapes and forms

Figure 4. Biophilic design process

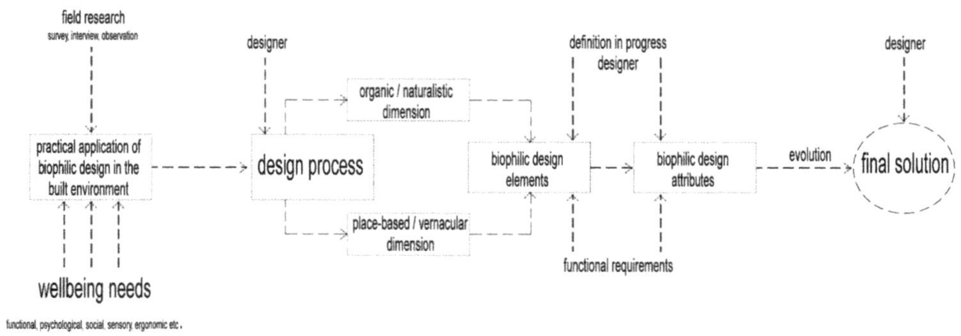

are botanical motifs, animal motifs, arches, vaults, domes and oval forms. Every design element finds concrete definitions with these attributes.

According to this detailed framework, biophilic design approach and its practical application is summarized in the figure below: two basic dimensions (naturalistic and vernacular), six biophilic design elements and 70 biophilic design attributes that focus on effects, details, concepts (Figure 4). This clear definition of design elements and attributes in Biophilia can be very helpful for the interior designer to understand and structure all the tools that he can use in order to promote wellbeing in interior space. In this sense the framework of Biophilia can be seen as a guide to concretize the relationship between human, nature and space.

User Centered Design

Human centered design and user-centered design are widely used replacing each other although it is possible to talk about a slight difference between the two terms. While human centred design is based on creating things deeply based on general natural characteristics and peculiarities of human psychology and perception, user-centered design is more focused on a deeper analysis of target audience. In architecture and interior design there are times that we design for a well-defined target audience and there are times that the target is not strictly defined. For example the definition of target is very different while designing a mass housing unit or a residence for a specific client. On the other hand both user-centered design and human centered design are based on the principle of answering to user needs on the ground of psychology, physiology, sociology and other sciences analyzing human life and interaction with the environment.

The term 'user-centered design' originated in Donald Norman's research laboratory at the University of California San Diego (UCSD) in the 1980s and became widely used after the publication of a co-authored book entitled: User-Centered System Design: New Perspectives on Human-Computer Interaction (Norman & Draper, 1986). He offers four basic suggestions on how a design should be:

- Make it easy to determine what actions are possible at any moment.
- Make things visible, including the conceptual model of the system, the alternative actions, and the results of actions.
- Make it easy to evaluate the current state of the system.
- Follow natural mappings between intentions and the required actions; between actions and the resulting effect; and between the information that is visible and the interpretation of the system state. (Norman, 1988, p.188)

While we consider living environments, well-being is strongly related to the needs and cultural backgrounds of users. In this sense, the approach of user-centered design is very significant for wellbeing research in interiors. If the users are already known, an in-depth study about their expectations and requirements is the basic input for user satisfaction. At this point we can say that user-centred design is an approach that can contribute to well-being of users as it is based on the physical and psychological needs of the users, enabling them to function at the highest level possible.

User-centred design is not a design style, but is an approach, a process for designing and developing buildings, products, and communities that is grounded in information about the people who will be using them—utilizing research findings and data on cognitive abilities, physical abilities and limitations, social needs, and task requirements in order to provide living-environment solutions that enable all users to function at their highest capacity—regardless of age or ability (Greenhouse, 2012). Some types of user-centered design consult users about their needs and involve them at specific times during the design process; typically during requirements gathering and usability testing. At the opposite end of the spectrum there are methods in which users have a deep impact on the design by being involved as partners with designers throughout the design process (Abras, Maloney-Krichmar, & Preece, 2004).

The user-centered design process begins with defining the design parameters according to user needs and well-being needs and the data collection process greatly depends on direct survey studies and interviews with the user. Then the data gathered from these studies are developed into design parameters. Through the design process the designer tries to develop design alternatives in order to test their efficiency with direct feedback from the user and works with these findings in order to redesign and develop better solutions that answer to the practical needs of users (Figure 5). The contribution of user through the design process can be regarded as an advantage that can also increase the efficiency of functional, ergonomic, sensory and psychological aspects related to well-being. On the other hand, user contribution should not be regarded as an added value but it should always be part of the design process while dealing with interior space. Otherwise design solutions can only be representatives of designer's ideas, but not those of real users. So, interior design for well-being should always consider user contribution as a part of the data collection and evaluation process (Minucciani, Saglar Onay, 2018).

Inclusive Design

Inclusive design has emerged in the UK and other countries (Coleman, 1994) from collaborations between industry, designers, researchers and educators. It constitutes a framework and growing body of practice within which design practitioners can understand and respond to the needs of diverse users, with

Figure 5. Human centered design process

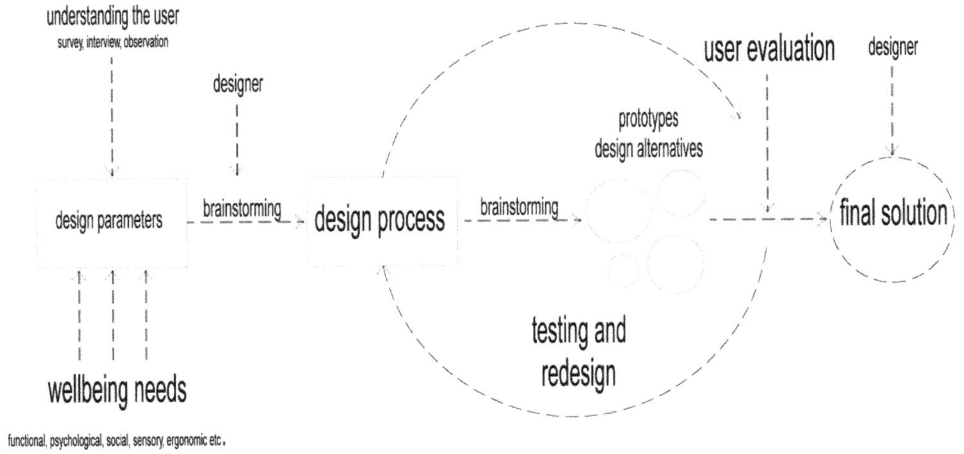

the ultimate inspiration of developing products and services that can meet the needs of the whole population (Clarkson, Coleman, Keates, Lebbon, 2003). On the other hand, inclusive design is mostly used and discussed for design products but not for architecture and interior design. Because for environmental design disciplines designing for the specific user groups is already accepted as a design priority. But this necessitates an in-depth understanding of well-being requirements of these different groups, away from predeterminations. The conception of inclusive interior design is to be handled with all the specific requirements that built up well-being taking into account the changing tendencies of requirements related to different age groups. According to Imrie and Hall (2001), in inclusive design, a very important concept is to understand why design solutions can not be useful for some parts of the society, what is special about them and why they don't fulfil the needs of this specific group. Understanding the missing points leads to a better understanding of how the design can become better. In this sense inclusive design is concerned with meaning and context, participative, human oriented, user re-defined to include users, low cost, democratic and heterogeneous (Imrie, Hall, 2001).

The Royal National Institute for the Blind (RNIB, 1995:13) in the UK noted that inclusive design does not look for the lowest common denominator, nor does it attempt to reconcile the often-conflicting needs of every possible minority group in society. Rather, by considering many varieties of special needs, inclusive design tries to break down unnecessary barriers and exclusiveness. Hatch (1984) suggests, inclusive design can, in part, be defined by 'users having the ability to take control of their environments'. Likewise, Towers (1995:172) notes that 'participation can be justified on the grounds that people have a right to greater control over their environment', while, for others, the absence of participatory democracy in development and design processes is a form of environmental injustice (Gleeson, 1999; Hester Jr., 1987).

The inclusive design process for interiors begins with understanding the needs of different user groups and the ways they differ from the others. Then design alternatives are to be developed in order to eliminate the obstacles for all possible user profiles and to include also positive attractions for all. On the other hand the inclusiveness of design proposals are to be evaluated by observations and evaluations of different user groups to understand the missing points and eliminate them (Figure 6).

Figure 6. Inclusive design process

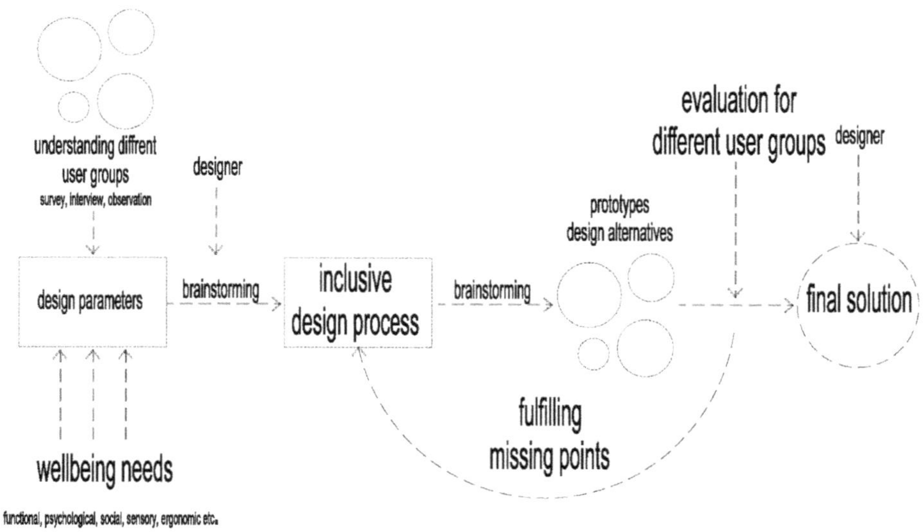

Supportive Design

Supportive design is a design approach that has been discussed especially in healthcare design in order create environments that can reduce stress. In this sense the main objective of this approach is to minimize the negative effect caused by the state of not being healthy and to maximize the healing effect by positive distractions. Ulrich (1991), has defined the three main objectives of supportive design as not raising obstacles to coping with stress, designing to facilitate access or exposure to physical features and social situations that have stress reducing influences and including visitors and healthcare staff as well as patients in the target group. He underlines that healthcare environments can promote well-being if they are designed to foster sense of control, access to social support and access to positive distractions.

Sense of control is evaluated as a factor that considerably reduces stress as users can decide on aspects that can have effect on their well-being and they can be selective. In psychology, having control over outcomes is associated with better-reported health, fewer and less severe symptoms, faster recovery from illness and greater longevity (Lachman, 1986; Rodin, Timko & Harris, 1985; Lachman, Weaver, 1998) On the other hand, too much sense of control can also be damaging. Those with views of the world as controllable and predictable may be particularly vulnerable when faced with an uncontrollable events (Wortman, Sheedy, Gluhoski, Kessler, 1992). Ulrich (1991) discusses that in healthcare environments lack of control is a pervasive problem that increases stress and adversely affects well-being. Due to reduced physical capabilities patients' sense of control can be significantly reduced. The contribution of the physical environment can lead to significant changes on sense of control. Design approaches can give the opportunity to control privacy so that patients can have increased control on their interaction with the surroundings. On the other hand staff can find escape areas designed and located to avoid frequent, unnecessary interruptions (Ulrich, 1991).

Social support as the second important actor in supportive design is related to different aspects of spatial design such as interiors' potential to enhance social interaction in an appropriate way. Studies from environmental psychology make evident that devastating consequences can arise from failing to achieve a balance between privacy and social interaction. According to Halpern (1995), if the built environment can influence mental health, then its role as a source of stress and its influence over social networks are two channels of its influence. Applying this observation to an architectural analysis, two issues come into the spotlight: the extent to which the architecture can facilitate incidental social interaction and the extent to which the environment communicates that it is for the user that they are not somehow excluded from the social spaces provided (McLaughlan, 2018). McLaughlan (2018), also discusses the term psychological in terms of supportive design. A spatially focused definition of psychosocially supportive space should refer to the provision of spaces that maximize opportunities for patients to access social support from within the hospital, either from family members or from other patients. The increasing preference for single-occupancy rooms over multi-bed ward accommodation disrupts traditional social practices within the pediatric hospital that facilitated the development of peer relationships, both for patients and their caregivers (McLaughlan, 2018).

The third key element in supportive design is positive distraction leading to positive feelings and moving the center of attention from sickness to the positive effect of certain attractions usually linked to nature and natural elements. Research in environmental psychology suggests that human well-being is usually fostered when physical surroundings provide a moderate degree of positive stimulation that is levels of stimulation that are neither too high nor too low (Wohlwill, 1968, Berlyne, 1971, Ulrich 1991). There has been growing interest among architects and other students of environmental phenomena in the question of whether or not complex, ambiguous, noisy, and otherwise stimulating urban environments are beneficial or harmful. A number of psychologists working in this area have advanced an optimal-level hypothesis, postulating an inverted-U shaped relationship between magnitude of stimulation along the dimensions considered above and the arousal value of, interest in, or preference for a given stimulus (Wohlwill, 2017). In this sense the stimulation levels offered by nature can be regarded as the most beneficial levels both for physical and psychological health. A growing amount of scientific evidence suggests that nature elements or views can be effective as stress-reducing, positive distractions that promote wellness in healthcare environments. In considering the needs of different types of users of healthcare facilities--patients, visitors, staff--it should be kept in mind that these groups sometimes have conflicting needs or orientations with respect to control, social support, and positive distractions. It is important for designers to recognize such differing orientations as potential sources of conflict and stress in health facilities (Schumaker and Pequegnat, 1989).

Becker, Sweeny, and Parsons (2008) have suggested that the atmospheric quality of a waiting room can influence patient perceptions of the quality of care they are receiving. In two waiting room studies that examined adult responses pre and post refurbishment, researchers concluded that the architectural treatment of an interior somehow buffers the negative effects of stress associated with a hospital visit (Becker et al., 2008; Leather, Beale, Santos, Watts, & Lee, 2003).

The design parameters in supportive design are to be determined after evaluating the disadvantages related to space and related to possible user groups. In the case of hospital environments there are some specific requirements related to type of illnesses and the ways they are to be cured. The main objective of this approach is to minimize the negative effect and to maximize the positive effect through the consideration of positive distractions, sense of control and social support as seen in Figure 7.

Figure 7. Supportive design process

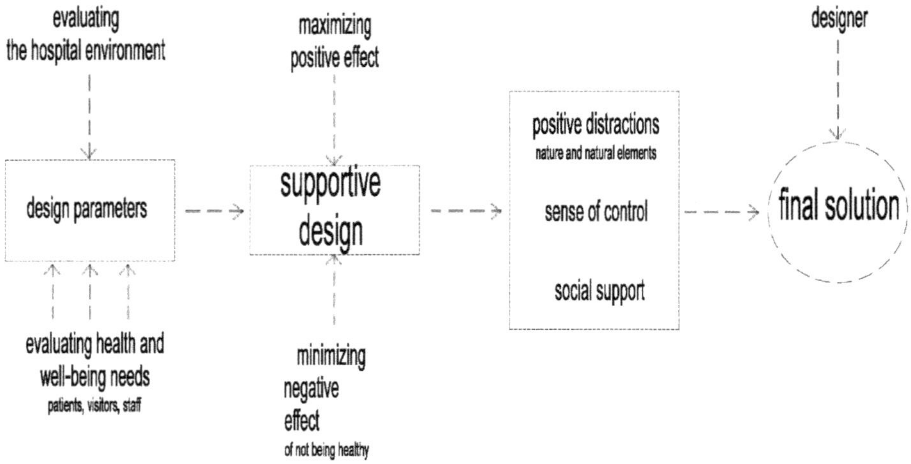

In a healthcare environment it is necessary to evaluate the spatial attributes that can be improved by understanding the necessities of treatment and therapy because in these interiors the priority is providing the necessary cure for the specific illnesses. The main parameters that can be evaluated can be spatial organisation, the evaluations of openings/views, the decisions about surface characteristics, materials and various objects that can evoke positive feelings.

The supportive design process for interiors can also be considered as an appropriate approach in order to rethink interiors with problematic conditions like poorly developed mass housing units or public interiors. In this sense reuse proposals can take the advantage of the key design elements of supportive design discussed above as positive distractions, sense of control and social support. The issue of positive distractions can be also associated with biophilic design elements and attributes. The designer can take the advantage of existing features in or around the interior environment or create new attractions that can be associated with the existing qualities of the interior environment.

Generative Design as an Emerging Design Approach

Generative Design can be regarded as one of the most important emerging design approaches. The added value of generative approach is the possibility to create a group of efficient solutions or alternatives rather than only one solution. In multi-criteria design problems where the needs, users and expectations can vary, generative approach can enrich the set of possible solutions. Thus, Generative systems offer a new insight into both the conceptualization of design processes and the study of design through dynamic processes and outputs. In literature genetic algorithms have been used in the field of design, optimization of designs, spatial arrangement and architectural form search. (Gu, Singh and Merrick,2010). McCormack, Dorin and Innocent (2004), define the properties of generative Design as follows:

- The ability to generate complexity
- The complex and interconnected relationship between organism and environment.
- The ability to self-maintain and self-repair
- The ability to generate novel structures, behaviors, outcomes or relationships

On the other hand the process of generative design is characterized by four main steps (McCormack, Dorin and Innocent 2004):

- Establishing requirements
- Designing alternatives
- Prototyping
- Evaluating

Generative Systems often use Genetic Algorithms as the search and optimization engine. Genetic algorithms are a generative design method inspired by the evolutionary process of nature and they simulate a long time natural selection in a short time. Evolutionary systems are based on simulating the process of natural selection and reproduction on a computer (McCormack, Dorin and Innocent, 2004). Initially a 'population' of potential designs is generated with a random set of parameters. This random population may be displayed visually to the designer. The designer's aesthetic sense then determines the 'fittest' designs of those displayed, and these are 'bred' with one another to produce a new population of designs that inherit the traits of their successful parent (Dorin 2001). In this sense, both computer and designer have an important role in generative approach. While the computer generates alternatives, the designer guides the selection giving the direction towards more successful generations. According to Rosenman (1997) and Carranza (2005) generative systems help to achieve not optimal, but satisfactory results according to the criteria determined by the designer (Minucciani, Saglar Onay, 2018).

A common way of using generative design approach is to code fitness functions into the evolutionary system. The designer can simply code the properties that he desires or he can define some restrictions, which can guide the evolution process. According to McCormack and Dorin (2001), design using generative methods involves the creation and modification of rules or systems that interact to generate the finished design autonomously. Hence, the designer does not directly manipulate the produced artifact, rather the rules and systems involved in the artefact's production. Correspondingly Meibodi (2016) discusses that the design process is automated and carried out by the computer, while the designer becomes a part of the productive apparatus, defining links and assigning directions of links between mediating artefacts (Minucciani, Saglar Onay, 2018).

According to Caldas (2001), the generative design approach is open-ended, as it leaves the final decision-making to the designer. The generative system is then used to generate whole building geometries, departing from abstract relationships between design elements and using adaptation to evolve architectural form. The shape-generation experiments are performed for distinct geographic locations, testing the algorithm's ability to adapt buildings shape to different environments (Minucciani, Saglar Onay, 2018).

Akella (2018) discusses that using artificial intelligence (AI) software, generative design enables engineers to create thousands of design options by simply defining their design problem - inputting design parameters (such as materials, size, weight, strength, manufacturing methods, and cost constraints) into generative design software and the software explores all the possible combinations of a solution, quickly generating hundreds or even thousands of design options. With the emerging production methods such as 3d printing, generative design gives the possibility to create extremely strong, efficient and lightweight shapes. The interior partition for Airplane manufacturer Airbus was designed to be lighter in order to reduce overall weight of the plane leading to the reduction of fuel consumed and carbon dioxide emitted when applied across its fleet of planes. The resulting partition was 45 percent lighter than the previous one as all the unnecessary inner filling material was removed (Figure 8) (Minucciani, Saglar Onay, 2018).

Figure 8. Airbus bionic partition
(Akella, 2018)

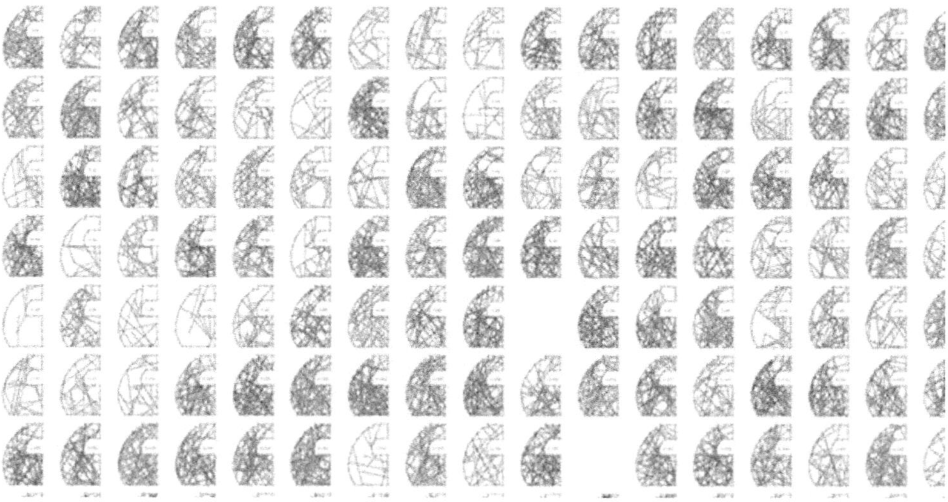

Generative design is not limited to product development but it can also be applied to -living environments. Saglar Onay, Garip and Garip used generative design in order to create flexible interior design solutions for standard mass housing units. The design team needed to create a pool of interior solutions that could answer to the needs of different families with different backgrounds. In the first stage, the apartments' three-dimensional digital model was created. Afterwards some fitness functions were defined in order to place the furnishings on the 3D grid. The genetic algorithm produced design alternatives by simultaneously considering each fitness function during the installation of furnishings. Moreover some restrictions have been made to prevent unrealistic alternatives during the operation. During the study, alternatives were dealt with in 3 steps. The genetic algorithm worked with the Grasshopper Octopus plug-in to produce alternatives. In the first phase, the alternatives that did not meet the fitness functions described as "unsuccessful" were eliminated. Ideal alternatives developed during the first phase were left to evolve and produce "successful" alternatives. At the latest stage, a certain number of alternatives selected by the designer formed the "most successful" solution set. In Figure 9, the alternatives produced by the

Figure 9. Fitness functions represented as cubes In the "Pareto Front" graphic (Saglar Onay, Garip and Garip, 2017).

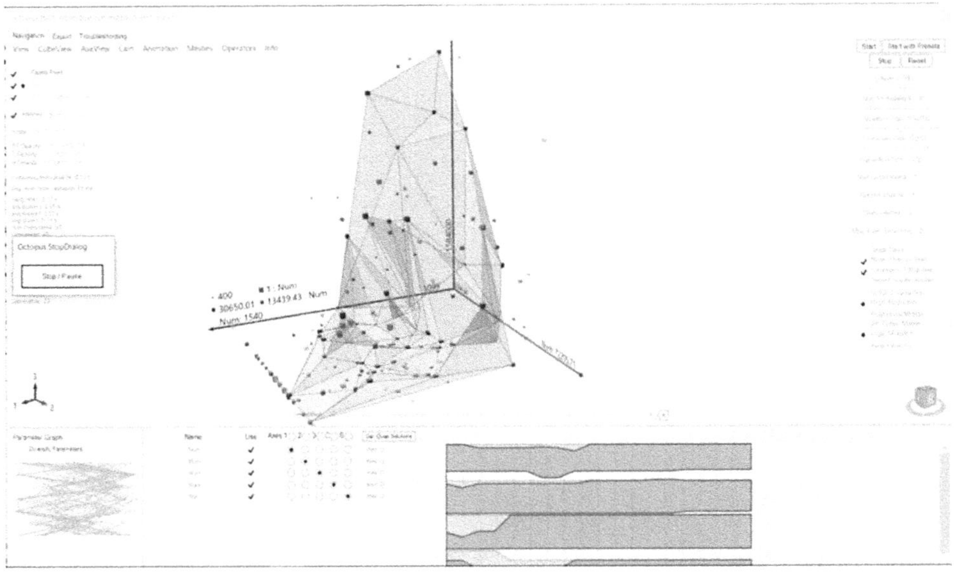

Octopus extension are represented as cubes in the graphic "Pareto Front" (Saglar Onay, Garip and Garip, 2017).

The X, Y, and Z axes in the Pareto Front graph indicate which alternatives better match with which fitness function. At this point, it can be said that the centrally located cubes are design solutions optimized for multiple fitness functions. The red cubes represent the alternatives that are eliminated as unsuccessful and the gray area indicates the range of evolutions. As satisfactory results begin to be produced, the gray area approaches the center. Figure 12 shows the process applied to one of the bedrooms. The matrix is created by selecting three alternatives from the designs evolving 5,100, 200 and 300 times respectively (Figure 10). The furnishings in the matrix are represented as solid geometries.

The potential of Generative Design in solving multi criteria design problems is to be evaluated and tested in order to generate well-being in interior living environments. In this case the determination of design parameters plays a very important role. Therefore the parameters to be coded in the generative system can be defined in order to fulfill well-being needs outlined in the well-being framework (Figure 2) together with the data collected from field studies. For various well-being requirements such as functional, ergonomic, sensory etc, the well-being parameters are to be determined. This can be achieved by the

Figure 10. A matrix created by selecting three alternatives from the designs evolving 5,100, 200 and 300 times respectively
(Saglar Onay, Garip and Garip, 2017).

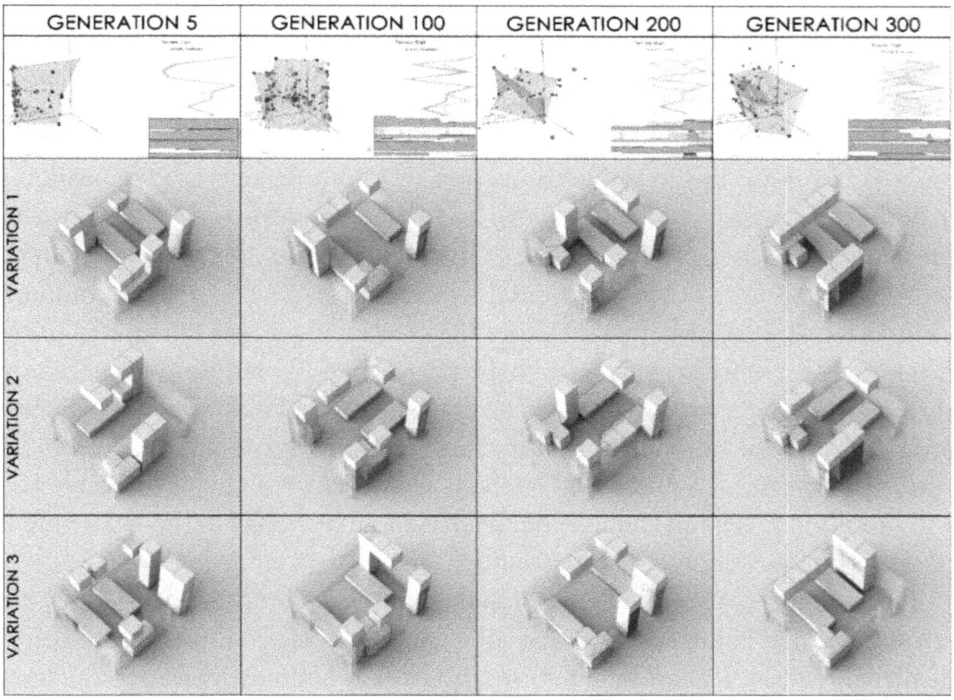

Figure 11. The use of Generative Design to generate well-being in interiors

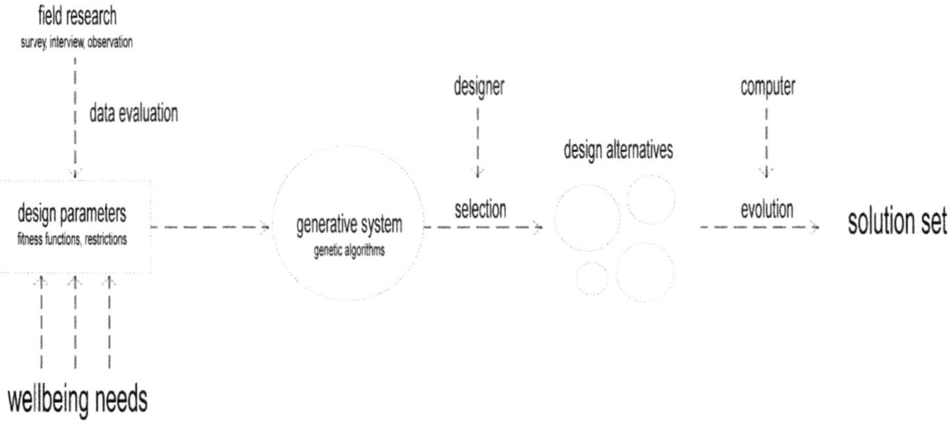

definition of appropriate fitness functions that fulfill well-being criteria as well as restrictions that will eliminate generation of possible inappropriate alternatives (Figure 11).

EVALUATION OF DESIGN APPROACHES

As we have seen in the previous section different design approaches have different priorities. Many of them have been developed as a response to problematic environmental situations. Inclusive design, in this sense, can be seen as a response to environmental injustice. It tries to create a common ground for all user groups that share the environment and the contribution becomes even more significant considering especially public spaces. Likely supportive design aims to eliminate the negative effect of having health problems and tries to give relief especially by developing the psychological state of users. Approaches based on the positive effect of nature like biophilic and biomimetic design can also be regarded as a response to the detachment of our living environments from nature.

Design approaches with particular focus on natural world have the advantage of creating strong ties with context, which eliminate the sense of placelessness. If elaborated correctly, natural sources as water, air, light etc. have also great potential to stimulate positive sensory responses as well as psychological ones. While biophilic design focuses on place based and naturalistic dimensions of the surrounding environment, biomimetic design focuses on local natural systems to fulfill the design requirements. Human centered design becomes more advantageous for answering to functional requirements. But the prototyping and testing processes also give the possibility to enhance ergonomic, psychological social and even sensory aspects with the feedback from users (Minucciani, Saglar Onay, 2018).

Flexibility becomes an important design issue while considering the subjective character of well-being in interiors. Therefore especially interiors with multiple or changing users need to be considered with a design approach that can lead to multiple solutions that can be adapted to changing users and preferences. While compared with conventional design approaches, generative design comes one step forward with its potential to create flexibility. In this sense generative approach can create multiple solutions or alternatives to fulfill different functional needs for different users with different sensory and psychological responses. But this flexibility greatly depends on the appropriate definition and hierarchy of design parameters as well as the correct selection of efficient solutions (Minucciani, Saglar Onay, 2018).

Table 1 underlines the advantages of design approaches discussed above according to the different spatial requirements of well-being addressed in Chapter 5; A design framework for well-being in interiors. The framework proposal, consisting of contextual, functional, psychological, social, ergonomic, aesthetic and sensory requirements as basic design criteria, aims to analyse the multi-layered structure of well-being related to space. Connection to context has an important role as the factor that makes every interior unique with its location and relationship to the rest of the world. Functional requirements are concerned with creating spatial solutions that support human activities. Psychological requirements are more indirect as

Table 1. Advantages of design approaches according to well-being requirements

	Connection to Context	**Functional**	**Social**	**Sensory**	**Ergonomic**	**Psychological**
Human centered design		strong connection to user needs	advantage of testing through prototype *observation *interview			
Biophilic design	strong connection to natural environment *place based *naturalistic	expression of a biological need	promotion of human affinity to natural world	advantage of natural stimulators	advantage of vernacular patterns	benefits of connectedness with nature
biomimetic design	strong connection to natural environment *observation of natural world	solutions according to natural processes and systems	sustainable solutions	advantage of natural processes	advantage of natural solutions	peaceful coexistence with the natural world
inclusive design	Elimination of all contextual obstacles for inclusiveness	Understanding different functional needs of users	Evaluation of different social groups and togetherness		Consideration of ergonomic capabilities of different user groups	Elimination of environmental injustice
supportive design	Use of nature contact for positive distraction		Focus on social support for healing	Focus on sense of control for healing		Positive attractions and social support for psychological healing
generative design	depends on defined parameters	flexibility multi-functional solutions	flexibility multi-user solutions			

they deal with the outcomes of the interaction between human and space. Social requirements deal with the spatial opportunities for the level of interaction between people. Sensory requirements are concerned with both quantitative factors like heating, illumination, acoustics, sound control etc. and their reflections on human senses. Aesthetic requirements are also important as they refer to the integrity and harmony of all spatial elements that can be pleasing to human senses and can promote well-being. Ergonomic requirements are more concrete and they refer to the relationship between all interior components and human body and capabilities. So, in Table 1, it is possible to see that different approaches generate different solutions for each requirement in the well-being framework. It is important to note that the consideration of all requirements is necessary in order to achieve a holistic design approach that leads to well-being. On the other hand approaches can give more importance to certain requirements according to design priorities and problems to be solved. For example biophilic and biomimetic design gives more importance to connection to natural context and handles all other requirements in connection to natural aspects. On the other hand supportive design focuses on psychological requirements and considers other requirements as contributors to the psychological state of users.

FUTURE RESEARCH DIRECTIONS

Further research needs to exemplify the use of different design approaches and evaluate their efficiency through real-world design problems. Among all discussed approaches, generative design can be regarded as the most encouraging one because of its potential to solve complex design problems by proposing multiple and flexible solutions.

CONCLUSION

Undoubtedly many different design approaches can be used to achieve well-being in interiors. As the described examples show, a really holistic approach to interior design is very challenging to achieve, and research can only be experimental. But every design approach handles the issue of design from different perspectives and has the potential to contribute to the understanding of well-being related to space. They all have different focuses to improve the spatial quality of our living environments. In this sense inclusive and supportive design focus on eliminating effects of problematic conditions. Approaches like biophilic and biomimetic design focus on the positive effect of nature trying to take advantage of natural stimulators and processes. Human centered design focuses on testing through prototypes and generative design takes the advantage of computer-designer interaction in order to create multiple design solutions. Therefore, it is impossible to classify existing design approaches according to their capacities to promote well-being as every design problem has its own priorities.

An important issue in design for wellbeing is the balance between freedom and determination. Design needs to address appropriate ways of behaving without blocking people's ability to create their own answers to the question of the good life. Therefore an appropriate understanding of freedom is a necessary ethical complement of all design activities including interior design. In this sense design approaches need to handle the issue of wellbeing in a flexible way and should avoid defining strict design rules for wellbeing.

Well-being in interiors is obviously a difficult and multi-criteria design task. Therefore design methodologies are very important in order to overcome the complexity of design problems. In this sense we can say that every design problem can develop its own design approach, as they are different especially related to context and users. A comprehensive understanding on how the environment and people relate to each other may make the difference between a single solution with a blind angle, and a complex solution where different scales of environmental design get into each other in continuation with a interdisciplinary understanding.

REFERENCES

Abras, C., Maloney-Krichmar, D., & Preece, J. (2004). User-centered design. In W. Bainbridge (Ed.), *Encyclopedia of Human-Computer Interaction*. Sage Publications.

Akella, R. (2018). *What generative design is and why it's the future of manufacturing*. Retrieved from http://www.newequipment.com/research-and-development

Barker, P., Barrick, J., & Wilson, R. (1995). *Building sight: A handbook of building and interior design solutions to include the needs of visually impaired people*. The Royal National Institute for the Blind.

Becker, F., & Douglass, S. (2008). The ecology of the patient visit: Physical attractiveness, waiting times, and perceived quality of care. *The Journal of Ambulatory Care Management*, *31*(2), 128–141. doi:10.1097/01.JAC.0000314703.34795.44 PMID:18360174

Benyus, J. (1997). *Biomimicry. Innovation inspired by Nature*. Harper Collins.

Benyus, J. (2008). A good place to settle: biomimicry, biophilia, and the return of nature's inspiration to architecture. In *Biophilic Design. The theory, science, and practice of bringing buildings to life*. John Wiley & Sons.

Berlyne, D. E. (1971). *Aesthetics and psychology*. Appleton Century Crofts.

Caldas, L. G. (1999). *An evolution-based generative design system: Using adaptation to shape architectural form* (PhD Thesis). Massachusetts Institute of Technology.

Carranza, P. M. (2001). *Self-design and ontogenic evolution*. Interactive Institue.

Clarkson, J., Coleman, R., Keates, S., & Lebbon, C. (2003). *Inclusive design: Design for the whole population*. Springer Verlag. doi:10.1007/978-1-4471-0001-0

Coleman, J. S. (1994). *Foundations of social theory*. The Belknap Press of Harward University Press.

Cramer, J. S., & Browning, W. D. (2008). Transforming building practices through biophilic design. In S. F. Kellert, J. H. Heerwagen, & M. L. Mador (Eds.), *Biophilic Design* (pp. 335–346). Wiley.

Desmet, P. M. A., & Pohlmeyer, A. E. (2013). Positive design: An introduction to design for subjective well-being. *International Journal of Design*, *7*(3), 5–19.

Dorin, A. (2001). Aesthetic fitness and artificial evolution for the selection of imagery from the mythical infinite library. In *Proceedings of the 6th European Conference on Advances in artificial life*, (pp. 659-668). Prague: Springer-Verlag. 10.1007/3-540-44811-X_76

Dorrestijn, S., & Verbeek, P. P. (2013). Technology, well-being, and freedom: The legacy of utopian design. *International Journal of Design*, *7*(3), 45–56.

El-Zeiny, R. M. A. (2012). Biomimicry as a problem solving methodology in interior architecture. *Procedia: Social and Behavioral Sciences*, *50*, 502–512. doi:10.1016/j.sbspro.2012.08.054

Fogg, B. J. (2003). *Persuasive technology: Using computers to change what we think and do*. Morgan Kaufmann. doi:10.1016/B978-155860643-2/50011-1

Gleeson, B. (1999). *Geographies of disability*. Routledge. doi:10.4324/9780203266182

Greenhouse, E. S. (2012). *Human centered design, Livable New York Resource Manual*. Retrieved from http://www.aging.ny.gov/LivableNY/ResourceManual/Index.cfm

Halpern, D. (1995). *Mental health and the built environment: More than bricks and mortar?* Taylor and Francis.

Hatch, R. (1984). *The scope of social architecture*. Van Nostrand Reinhold.

Heerwagen, J. H. (2006). Investing in people: The social benefits of sustainable design. Proceedings of Rethinking Sustainable Construction.

Hester, R. Jr. (1987). Participatory design and environmental justice. *Journal of Architectural and Planning Research*, *4*(4), 289–300.

Imrie, R., & Hall, P. (2001). *Inclusive design: Designing and developing accessible environments*. Spon Press.

Kellert, R. (2005). *Building for life: designing and understanding the human-nature connection*. Island Press.

Kellert, R. (2008). *Biophilic design: The theory, science and practice of bringing buildings to life*. John Wiley & Sons.

Kellert, R., & Wilson, E. O. (1993). *The biophilia hypothesis*. Island Press.

Kennedy, S. (2004). Biomimicry/biomimetics. General principles and practical examples, *The Science Creative Quarterly, 6*.

Lachman, M. E. (1986). Locus of control and aging research: A case for multidimensional and domain specific assessment. *Psychology and Aging*, *1*(1), 34–40. doi:10.1037/0882-7974.1.1.34 PMID:3267376

Lachman, M. E., & Weaver, S. L. (1998). The sense of control as a moderator of social class differences in health and weel-being. *Journal of Personality and Social Psychology*, *74*(3), 763–773. doi:10.1037/0022-3514.74.3.763 PMID:9523418

Lyubomirsky, S., Sheldon, K. M., & Schkade, D. (2005). Pursuing happiness: The architecture of sustainable change. *Review of General Psychology*, *9*(2), 111–131. doi:10.1037/1089-2680.9.2.111

McCormack, J., & Dorin, A. (2001). Art, emergence and the computational sublime. In A. Dorin (Ed.), *Second Iteration: a conference on generative systems in the electronic arts* (pp. 67–81). CEMA.

McCormack, J., Dorin, A., & Innocent, T. (2004). Generative design: a paradigm for design research. In *Proceedings of Future ground*. Design Research Society.

McLaughlan, R. (2018). Psychosocially supportive design: The case for greater attention to social space within the pediatric hospital. *HERD: Health Environments Research & Design Journal*, *11*(2), 151–162. doi:10.1177/1937586717731739 PMID:29022369

Meibodi, M. A. (2016). *Generative design exploration: Computation and material practice*. TRITA-ARK Academic Publishing.

Minucciani, V., & Saglar Onay, N. (2018). Evaluation of design approaches for well-being in interiors. *Journal of Engineering and Architecture*, *6*(1), 112–122.

Naci, H., & Ioannidis, J. P. A. (2015). Evaluation of wellness determinants and interventions by citizen scientists. *Journal of the American Medical Association*, *314*(2), 121–122. doi:10.1001/jama.2015.6160 PMID:26068643

Norman, D. (1988). *The design of everyday things*. Doubleday.

Pawlyn, M. (2011). *Biomimicry in architecture*. R IBA Publishing.

Rodin, J., Timko, C., & Harris, S. (1985). The construct of control: Biological and psychological correlates. In C. Eisdorfer, M. P. Lawton, & G. L. Maddox (Eds.), *Annual Review of Gerontology and Geriatrics* (pp. 3–55). Springer.

Rosenman, M. A. (1997). The generation of form using an evolutionary approach. In D. Dasgupta & Z. Michalewicz (Eds.), *Evolutionary algorithms in engineering applications* (pp. 69–85). Springer. doi:10.1007/978-3-662-03423-1_4

Ryan, C. O., Browning, W. D., Clancy, J. O., Andrews, S. L., & Kallianpurkar, N. B. (2014). Biophilic design patterns: Emerging nature-based parameters for health and well-being in the built environment. *Archnet-IJAR*, *8*(2), 62–76. doi:10.26687/archnet-ijar.v8i2.436

Saglar Onay, N., Garip, B. & Garip, E. (2017). A flexible user centered design model for social housing units. *Tafter Journal, 97.*

Saglar Onay, N., & Minucciani, V. (2018). Well-being framework as a contributor to sustainability. *WIT Transactions on Ecology and the Environment, 217*, 699–710. doi:10.2495/SDP180591

Shumaker, S. A., & Pequegnat, W. (1989). Hospital design, health providers, and the delivery of effective health care. In E. H. Zube & G. T. Moore (Eds.), *Advances in Environment, Behaviour and Design, 2* (pp. 161–199). Plenum. doi:10.1007/978-1-4613-0717-4_6

Thaler, R. H., & Sunstein, C. R. (2008). *Nudge: Improving decisions about health, wealth, and happiness.* Yale University Press.

Tov, W., & Diener, E. (2007). Culture and subjective well-being. In S. Kitayama & D. Cohen (Eds.), *Handbook of cultural psychology* (pp. 691–713). Guilford.

Towers, G. (1995). *Building democracy: Community and architecture in the Inner Cities.* UCL Press.

Ulrich, R. S. (1991). Effects of interior design on wellness: Theory and recent scientific research. *National Symposium on Health Care Interior Design (U.S.), 3*, 97-109.

Vincent, J. (1990). *Structural biomaterials.* Princeton University Press.

Wilson, E. O. (1984). *Biophilia.* Harvard University Press.

Wohlwill, J. F. (1968). The physical environment: A problem for a psychology of stimulation. *The Journal of Social Issues*, *22*(4), 29–38. doi:10.1111/j.1540-4560.1966.tb00546.x

Wohlwill, J. F. (2017). The physical environment: A problem for a psychology of stimulation. In R. Gutman (Ed.), *People and Buildings.* Routledge. doi:10.4324/9781315126197-8

Wortman, C. B., Sheedy, C., Gluhoski, V., & Kessler, R. (1992). Stress, coping and health: conceptual issues and directions for the future research. In H. S. Friedman (Ed.), *Hostility, Coping and Health* (pp. 227–256). American Psychological Association. doi:10.1037/10105-016

KEY TERMS AND DEFINITIONS

Design Approach: The way the design input is evaluated and processed.

Genetic Algorithms: A generative design method inspired by the evolutionary process of nature that simulate a long-time natural selection in a short time.

Chapter 5
A Framework for Well-Being in Interiors

ABSTRACT

This chapter aims to propose an environmental design framework in order to contribute to sustainability with well-being criteria focusing on human factors. Designers need to handle the issue of well-being with a methodological approach, as it is very difficult to achieve with an intuitive attitude. There are several requirements that need to be fulfilled in order to create the environment that can promote user well-being. The framework proposal, consisting of contextual, functional, psychological, social, ergonomic, aesthetic, and sensory requirements as basic design criteria, aims to support both theoretical and practical activities regarding well-being in all living environments, as a crucial component of sustainability. In this sense, this chapter discusses all the components of the well-being framework and evaluates the effect of cultural differences on the hierarchy of these requirements.

INTRODUCTION

While considering major factors that contribute to well-being, it is important to differentiate those that can be controlled. Among controllable factors, the natural and built environment must be considered as a very important issue strongly connected to positive cognitive, behavioural, and goal-based activities. Therefore, space can be regarded as the most important concrete aspect that can be controlled by the designer and can directly affect human

DOI: 10.4018/978-1-7998-4231-6.ch005

well-being. This is also due to the fact that people pass an important part of their time indoors in designed environments. Especially home, school and working spaces must be regarded as living environments, which have significant role on our well-being. In spite of their great impact on our daily lives, research about well-being in living environments is very limited. In this sense this chapter aims to discuss the controllable factors in our living environments through a framework focusing on well-being in interior space.

Considering studies about well-being, human needs and environment, the research of Biologist Stephen Boyden, can be regarded as one of the earliest and comprehensive studies that discuss the environmental and spatial aspects of well-being. In their article Human ecology and the quality of life, Boyden and Millar (1978) state that the study of the total environment is fundamentally similar to system ecology in biological science, and involves consideration of the components of the 'system as a whole' and their dynamic interrelationships. It differs from biological system ecology in that the systems under investigation contain, in addition to inorganic and biotic components and processes, the components and processes of human culture. The second orientation is referred to as 'human experience'. It is based on recognition of the fact that each individual human being experiences his own personal life conditions, which are influenced by the properties of the total environment, and which, in turn, are the main determinants of his level of health and well-being. Boyden, in his research in 1971, distinguishes between "survival needs" and "well-being" needs. Survival needs deal with aspects of the environment that directly affect human health, such as clear air and water, lack of pathogens or toxins, and opportunity for rest and sleep. Well-being needs, on the other hand, are more indirect in their locus of impact. These needs affect overall health through their relationship to fulfilment, quality of life, and psychological health. Where failure to satisfy survival needs may lead to serious illness or death, failure to satisfy the well-being needs produces the "grey life" of psychosocial maladjustment and stress related illnesses. Taken as a whole, the research by Boyden and others identifies well being needs that should be addressed in building design:

- Opportunity to engage in spontaneous social encounters.
- Opportunity for relaxation and psychological restoration.
- Opportunity for privacy and for movement between interaction and solitude, as desired.
- Opportunity for learning and information sharing.
- Opportunity for connection to the natural environment.

- Opportunity for regular exercise.
- Sound levels not much above or below that of nature.
- Meaningful change and sensory variability.
- An interesting visual environment with aesthetic integrity.
- Sense of social equity and respect.
- Ability to maintain and control personal comfort.
- Making sense of the environment.

In this research we can see that Boyden discusses the contextual, social, psychological, sensorial and aesthetic dimensions of environmental design in relation to human needs. He also underlines the importance of meaningful change and equilibrium between contrasting conditions like solitude and interaction. This can be interpreted as the need for spatial flexibility, which is mainly concerned with functional aspects of space. The research of Boyden tries to define design criteria and a structured framework for healthier living environments and underlines the effect of design on our well-being.

According to the report of the Green Building Council (2016) (Figure1), well-being in home environments should be handled in two basic steps as the interior home environment and the neighbourhood surrounding it. Regarding healthy homes, the report underlines the importance of quantitative factors including indoor air quality, thermal environment, moisture, light, noise and qualitative factors including functionality, internal layout, ease of use, privacy, control, views of nature, safety, flexibility and social interaction. Regarding healthy neighbourhood the report highlights quantitative factors as air pollution, water quality, waste, public spaces and density and qualitative factors as accessibility, connectivity, transport networks, local amenities, defensible space, sports and leisure. The report discusses that these factors affect the health and well-being of occupants in both subjective and objective ways.

THE WELL-BEING FRAMEWORK FOR INTERIORS

Well-being in interiors is extremely subjective and it is very difficult to measure with objective criteria. On the other hand, designers need to handle the issue of well-being with a methodological approach, as it is very difficult to achieve well-being with an intuitive approach. In this part of the book, the well-being issue in interiors is structured by the help of a framework that focuses on well-being requirements (Figure 2). These requirements and the framework in general differentiate from other studies especially with its focus on

Figure 1. Healthy home indicators and metrics
(Green Building Council, 2016)

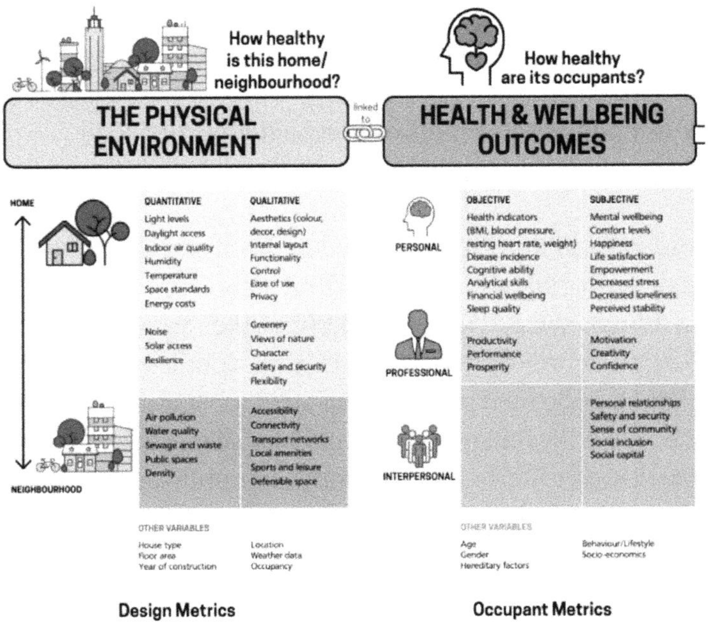

interiors and human needs. These requirements are also different than general requirements of design for well-being as they are strongly connected to space and context. Therefore, the well-being framework for interiors aims to define the spatial dimensions of well-being in order to evaluate the importance of every spatial aspect that has an impact on our well-being related to interior

Figure 2. Well-being Framework for Interiors

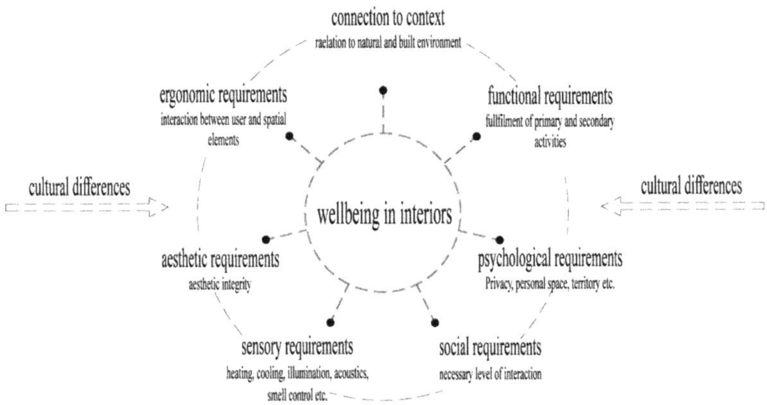

space. On the other hand the framework does not attempt to isolate these spatial aspects or requirements from each other as they are always in relation to each other and they are always subject to change in relation to the dynamics of everyday life. Moreover all these requirements are subject to great cultural differences regarding users with different backgrounds.

The well-being framework for interiors can be used both as an evaluation method for existing environments and a roadmap for designing new living environments that can promote well-being. The structure of the framework is based on contextual, functional, psychological, social, sensory, aesthetic and ergonomic requirements, which are always in relation to each other. Connection to context has an important role as the factor that makes every interior unique with its location and relationship to the rest of the world. Functional requirements are concerned with creating spatial solutions that support human activities. Psychological requirements are more indirect as they deal with the outcomes of the interaction between human and space. Social requirements deal with the spatial opportunities for the level of interaction between people. Sensory requirements are concerned with both quantitative factors like heating, illumination, acoustics, sound control etc. and their reflections on human senses. Aesthetic requirements are also important as they refer to the integrity and harmony of all spatial elements that can be pleasing to human senses and can promote well-being. Ergonomic requirements are more concrete and they refer to the relationship between all interior components and human body and capabilities.

EVALUATION OF WELL-BEING REQUIREMENTS IN INTERIOR SPACE

Connection to Context

Connection to context is the key element of the well-being framework. All other requirements have contextual dimensions and they are always in relation to contextual ties. When an architect is given a task to design a new building or to make an intervention on an existing one, the first thing to do is to understand the surrounding environment. In order to understand a building, it is necessary to grasp the exterior forces that shape it. Connection to context is the most important design input that creates architectural and spatial identity by differentiating buildings/spaces with context related data.

While we consider context of a building, it is important to discuss two basic dimensions, which are location and relationship to surroundings. While location is absolutely impossible to change, it provides the designer with the most valuable input for the design process like topographic features, climate etc. On the other hand, relationship to the surrounding environment is an aspect that the architect decides and for the interiors it has great importance as input in order to decide on the interior interventions. Regarding relationships, contextual ties can be discussed in terms of orientation, entrance, openings, surrounding buildings /spaces, infrastructure and natural aspects.

Contextual aspects of a building can be discussed also in terms of sense of place. According to Schulz (1980), place is something more than abstract location, a totality of concrete things having material substance, shape, texture and color. Together these things determine environmental character, which is the essence of place. A place is therefore a concrete total phenomenon, which we can not reduce to any of its properties such as relationships without losing its concrete nature out of sight. Sense of place is increasingly recognized as key to human well-being in social-ecological systems (Poe, Donatuto, Satterfield, 2016). It is in relation to those characteristics that make a place special or unique, as well as to those that foster a sense of authentic human attachment and belonging (Casey, 2001). Place attachment is also defined as a feeling of comfort, which the individual is benefited from in his or her living place (Brown, Perkins, & Brown, 2003). According to Wang et al. (2016), improving the place attachment leads to the promotion of social well-being and social cognitive function. Rollero & Piccoli (2010), also, conducted a study on 443 students at the University of Turin in Italy to assess the effect of the attachment to place on the social well-being and found that the attachment to place affects the social well-being (Rollero & De Piccoli, 2010). So we can say that correct relationships with context increase place attachment which is absolutely considered as a strong predictor for social well-being especially in older adults' population (Rollero & De Piccoli, 2010).

Just like architecture, the interior has to convey a message about its existence, about its context and content and it should reflect and respect sense of place. This is only possible if design components are integrated to each other with a design approach that considers the natural and built environment as a whole without ignoring human scale and senses. The geometry of the interior envelope as well as the geometry and the articulation of openings on the envelope create forces from inside to outside and from outside to inside. These borders help to control the interior's degree of interaction with its surroundings. The decisions about all these aspects define the way the interior interacts with the environment and makes the interior a part of genius loci mostly defined as the spirit of place.

Contact with nature is another issue to be considered in well-being and space studies. A growing body of knowledge supports the role of contact with nature in human well-being. Kellert (2016) notes that contact with nature has been found to enhance recovery from illness, that people living in proximity to open spaces report fewer health and social problems as well as superior quality of life and a stronger sense of place, that office settings with natural environmental features improve worker performance and motivation (Freeman, 2017). These studies provide scientific support for the assumption that contact with nature is critical to human well-being (Saglar Onay, Minucciani, 2018). It is very important that more attention is given to social and environmental contexts and their relations to well-being in interpersonal and social levels in future studies.

Functional Requirements

In general terms functional requirements can be regarded as the basic needs in living environments intended for a certain purpose and they are shaped by the main activities that take place in and around buildings. For a certain interior, the functional requirements may effect the organization of all interior elements such as furniture, lighting etc. The most efficient way of understanding functional requirements is to understand user's intended activities and their way of performing these activities. For multi purpose spaces, functional requirements need to be handled with a flexible approach, which give the possibility to use the same space for different purposes at different times or even for different user profiles. (Saglar Onay, Minucciani, 2018).

Function has been always considered as a key element in theory of architecture starting from Vitrivius to modernity. While for Vitrivius, utility was taken as one of the three main qualities (firmitas, utilitas, venustas) that a building should represent, for Sullivan, seen as the father of modernism, "form ever follows function". This phrase became a slogan of Modernist architects after the 1930s. The doctrine was taken to imply that decorative elements, which architects call "ornament", were superfluous in modern buildings and function was the main aspect to be considered to decide on the form and other aspects of a building. On the other hand there are also researchers that consider functional separations less decisive. For example, Herzberger (1991) notes that the differentiation of spaces according to their functions is just a choice and there is no need to make functional separations. In this sense a space, which is used for sleeping, can also be used for working. If necessary, an interior can be donated with interior elements that can support different activities during different times of the day, week and year. Such an understanding can be regarded as similar to the concept of traditional Turkish room, which could be used for all necessary activities performed by family members. Therefore, flexibility can be regarded as a major characteristic of the traditional Turkish house in various ways. The origin of the traditional Turkish house can be traced back to the Turkish way of life of the nomadic period, prior to settlement in Anatolia; it reflects effects of numerous cultures with which the Turkish people came into contact in both Central Asia and Anatolia (Turgut and Unugur, 1997). The concept of the extended family was an important factor in shaping the traditional Turkish house; fathers, mothers, children, sons and daughters-in-law lived under the same roof. The flexibility of the traditional Turkish room arises from the need to accommodate several nuclear families under the same roof. Therefore it becomes necessary to create a room layout that includes all private activities of a nuclear family. This understanding makes it possible to use less space for more activity and more people, which can be also suitable for today's public housing units.

Figure 3. Flexibility of "room" in traditional Turkish house
(Saglar Onay, Garip &Garip, 2019)

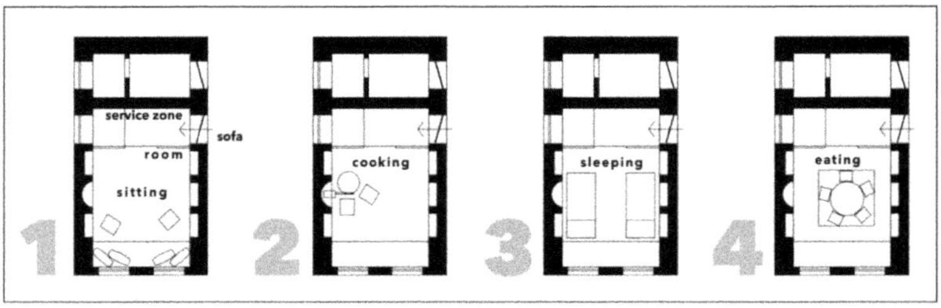

According to Charytonowicz (2000), our present needs and circumstance has lead to the creation within architecture of, amongst other things, an orientation towards a greater adaptability, both to the needs of all categories of mankind's efficiency and to the possibilities engendered by a more widely understood ecosystem of nature. Contemporary architecture has been becoming more of an "interdisciplinary engineering art" than a "fine art" and it is a form of the humanization of the technique and urbanized living environment.

Today with the emerging tendencies for reuse of existing buildings creates a different conception of function and functionality. In this sense the changes in society; ways of living and activities necessitates functional changes. Therefore other than potentials of existing buildings, it also becomes very important to consider both potential users and potential activities in order to decide on new interventions.

Ergonomic Requirements

The development of ergonomics followed two different approaches: one has its origin in health and safety, the other in human performance (Koningsveld, Settels, Pikaar, 2007). Regarding architectural space, ergonomy can also be discussed in these two basic aspects including both the appropriateness of all equipment and its efficiency for the fulfilment of human activity. Therefore, ergonomic requirements in interiors are related to the connection between human and space, mainly including physical and cognitive aspects, which help to fulfil a certain task. While physical ergonomics relate to physical activity mainly concerned with human anatomical, anthropometric, psychological and biomechanical characteristics, cognitive ergonomics focuses on mental processes regarding interactions among humans and surroundings (Salvendy, 2012). In this sense, user's capabilities and limitations in seeking to ensure the necessary task define ergonomic requirements. For Wilson (2000), the role of ergonomists is two-fold, albeit two parts that at their best make a seamless whole. First is to fundamentally understand purposive interactions between people and artefacts and especially to consider the capabilities, needs, desires and limitations of people in such interactions. In this role designers are scientists, a science that embraces qualitative enquiry in the field as much as controlled laboratory experimentation. The second role comprises a contribution to the design of interacting systems, maximising the capabilities, minimising the limitations, and trying to satisfy the needs and desires of the human race. In this role designers are craftspeople, using judgement, vision, experience and even trial and error to develop and test concepts and prototypes.

According to Pheasant (2003), human capacities and limitations can be understood through a repertoire of investigative methods. Among these two particular techniques deserve special attention: task analysis and user trial. Task analysis state what the user is actually going to do with the product/system/environment. An effective task analysis will clarify the overall goals of the project, establish the criteria that need to be met, point out the most likely areas of mismatch, and so on. On the other hand, user trial is an experimental investigation in which a sample of people tests a prototype version of the product under controlled conditions.

The most systematic research in the field of ergonomy related to space is macroergonomics, which is especially concentrated on work environments. Macroergonomics is particularly important as it tries to apply scientific knowledge about human capabilities, limitations, and other characteristics in order to improve the human condition, including health, safety, comfort, productivity, and quality of life (Human Factors and Ergonomics Society, 1998). Conceptually, macroergonomics may be defined as a top-down sociotechnical systems approach to the design of work systems, and the carry-through of the overall work system design to the design of the human-job, human-machine, and human-software interfaces (Hendrick, 1997; Hendrick & Kleiner, 2001). As a perspective, macroergonomics provides the ergonomist with an appreciation for the larger system. The goal of macroergonomics is to optimize the work system's design in terms of its sociotechnical system characteristics, and then carry the characteristics of the overall work system design down through to the design of individual jobs and human-machine and human-software interfaces to ensure a fully harmonized work system (Henrick, Kleiner, 2002). In this sense the approach of macroergonomics can be regarded as an important contribution to well-being research.

Interior design is a discipline that always has ergonomic concern as it is meant to deal with the environment that surrounds human body interacting with its both physically and cognitively. According to Charytonowicz (2000), the similarity of the main features and aims of activities which characterize both ergonomics and architecture justifies the need to include into the multidisciplinary circle of ergonomists also those architects who whilst creating the "material frameworks of man's life" still very often try "to create pure art" and forget about the "statistic man".

Psychological Requirements

Self-determination theorists (Deci, Ryan, 1985, Ryan, Deci, 2000), maintain that well-being hinges on the fulfilment of three basic innate psychological needs which are defined as:

- Competence: Seek to control the outcome and experience mastery

- Relatedness: desire to interact, be connected to, and experience caring for others
- Autonomy: desire to be causal agents of one's own life and act in harmony with one's integrated self.

Competence is defined as the ability to exert control over one's life, to cope with specific problems effectively, and to make changes to one's behavior and one's environment, as opposed to the mere ability to adjust or adapt to circumstances as they are (APA Dictionary of Psychology). Affirming, strengthening, or achieving a client's competence is often a basic goal in psychotherapy. Therefore competence can be associated with having control and being able to change the existing situation according to one's own needs and preferences. According to Keller (2016), humans are biologically prepared for developing competence in all possible environments with a universal repertoire of predispositions and propensities. Depending on the particular environmental information, some of these propensities are strengthened, whereas others disappear.

Relatedness is a feeling of connection with other people, often accompanied by affection, trust, and a sense of personal security (APA Dictionary of Psychology). So, relatedness can be also regarded as a social dimension of psychological needs. People need to care for other people and they need to be cared about by others. From a cultural, more precisely a societal or country point of view, the conceptions of individualism and collectivism have become pervasively prominent (Triandis 1995). In this conception, individualism relates to autonomy, whereas collectivism relates to relatedness (Kellert, 2016).

In general autonomy is defined as a state of independence and self-determination in an individual, a group, or a society. In self-determination theory more specifically, the experience of acting from choice, rather than feeling pressured to act. This form of autonomy is considered a fundamental psychological need that predicts well-being (APA Dictionary of Psychology). According to Keller (2016), autonomy/agency/mastery is referring to the intentional, mental world of wishes, desires, cognitions and control of individual options and choices. Relatedness on the other hand is assumed to cover the need for closeness with others and any social actions in the service of a social system. It is usually argued that these conceptions are valued and develop to different degrees depending on the cultural/geographical context like Euro-America or East Asia. However, both needs are equally important to everybody, yet they must manifest differently in different cultural/contextual environments.

According to Kupfer (1987), autonomy requires a conception of self for which privacy is indispensable. Privacy is a necessary condition for the development of an autonomous self and it is strongly linked to space. In this sense, environments should be responsive in terms of privacy and control. According to Namazian and Mehdipour [18] environments that emphasize only either very little interaction or a great deal of interaction are too static and will not be responsive to changing privacy needs so environmental designers should try to create environments that permit different degrees of control over contact with others.

Other than the three basic needs defined by Self-determination theorists, it is necessary to consider two other needs, which are very important in terms of well-being related to the environment. These needs are security and privacy. Security is defined as feelings of basic safety, assurance, and independence from alarm. In psychology, it is regarded to be engendered by such conditions as a comforting, approving support system of family and friends; progression of age-appropriate abilities and competencies; and occurrences that boost ego sturdiness (Psychology Dictionary). According to Schneier (2008), the psychology of security is intimately tied to how we think: both intellectually and emotionally. The feeling and reality of security are different, but they're closely related. The reality of security is mathematical, based on the probability of different risks and the effectiveness of different countermeasures. But the feeling is based not on probabilities and mathematical calculations, but on psychological reactions to both risks and counter measures. Therefore, the feeling of security is a feeling of being safe and free from worry.

Privacy is another important dimension of well-being related to space. It is both defined as the state in which an individual's or a group's desired level of social interaction is not exceeded and the right to control (psychologically and physically) others' access to one's personal world, such as by regulating others' input through use of physical or other barriers (e.g., doors, partitions) and by regulating one's own output in communication with others (APA Dictionary of Psychology).

All the psychological needs discussed above have great importance related to living environments. Privacy can be defined as having the desired degree of control over contact with others. Security can be defined as being and feeling free from potential harm. Competence related to living environment can be defined as having control over the environment and having the ability to change it if there is need. Relatedness can be interpreted as being connected to other people and Autonomy can be evaluated as being free to act as desired for one's own life and act in harmony with one's integrated self. Regarding the survey study about psychological requirements, the hierarchy of the importance these 5 needs has been evaluated by giving the simple definitions above (Table 6). This was very important in order to avoid misunderstanding about these concepts.

Social Requirements

As people live in community, well-being of individuals is related to well-being of others and the interaction between individuals and groups. Keyes (1998) defines social well-being as an individual's self report of the quality of his or her relationship with other people, the neighbourhood and the community.

According to Keyes and Shapiro (2004), social well-being has 5 dimensions including social integration (the quality of one's relationship to society), social contribution (evaluation of one's value to society), social coherence (perception of the quality, organization and operation of the social world), social actualization (evaluation of the potential and the trajectory of society) and social acceptance (construal of society through the character and qualities of other people as a generalized category). Although they seem unrelated to space, our activities promoting social well-being are directly affected by spatial properties. The spatial aspects of social well-being are related to qualities that enrich social life. In this sense, buildings should provide the chance for social encounters at desired level for everyone. Moreover they should support and encourage social activities creating a balance between public, semi-public and private places. The quantity and quality of social connections (e.g. talking and listening to family or strangers) correlate with reported well-being as well as physical health. The domestic environment provides the context for this, and the design of our homes and neighbourhoods can facilitate positive social interactions (Green Building Council, 2016).

Collective activities can be good supporters of social gatherings and increase the positive effect of getting together and sharing. For example in many cultures eating together is regarded as an activity that can increase happiness. According to Yiengprugsawan and others (2015), having a meal is not only important for nutritional and health outcomes; it is also a vital part of daily social interaction. Their study provides empirical evidence from a non-Western setting (Thailand) that sharing meals could contribute to increasing happiness.

Sensory Requirements

Sensory requirements are extremely effective on well-being. Ventilation, lighting, heating, cooling and acoustics are issues to be handled with special expertise. Interior designers need to pay attention both on ways to fulfil these requirements and their effects on human senses. Pallasma (2005) argues that every touching experience of architecture is multi-sensory; qualities of space, matter and scale are measured equally by the eye, ear, nose, skin, tongue, skeleton and muscle. Relatedly Zumthor (2006) states that we perceive atmosphere through our emotional sensibility, which is a form of perception that works incredibly quickly. We are capable of immediate appreciation, of a spontaneous emotional response. So atmosphere can be regarded as the essence of space, which we perceive through spatial experience.

According to Unwin (2009), in our experience of architecture, basic architectural elements and the places they identify are modified by light, color, sound, temperature, air movement, smell and the qualities and textures of the materials used. Possible configurations of basic and modifying elements are probably infinite. Therefore, as time goes on spatial experience and spatial atmosphere change continuously with the modifying elements of architecture. Sometimes these elements can be much more important than the physical definition of space as they stimulate all our senses. So, we need to take into account the effects of all these modifying elements in order to understand how they change (Saglar Onay, Minucciani, 2018).

A key element of achieving mental well-being is the element of control. When people feel in control of their surroundings and environment, they feel empowerment and stability (Shapiro, 1999), (Taylor et al, 1999). Sensory requirements tend to be controllable through physical arrangements related to thermal environment, acoustics, natural and artificial lighting, humidity and air quality. On the other hand, the effect of these arrangements might be different from person to person depending on personal attributes, activities and sensitivities. Among environmental forces, light and sound contribute significantly, individually and combined, towards the regulation of physio-psychological well-being, being most influential to perception and behavior.

Light is essential for well-being. On an emotional level, people enjoy and feel a sense of well-being in daylight and sunlight. Natural light makes us aware of the passage of time: the specific mechanisms of this interaction are not entirely understood, but our bodies are naturally in tune with external light levels and characteristics. As well as the photoreceptors responsible for vision, the eye contains sensors, which detect the blueness of daylight. Therefore, the colour and intensity of the light that we experience influences the secretion of hormones, including melatonin and serotonin, which regulate our circadian rhythms (Green Building Council, 2016). On the other hand, artificial lighting is very much driven by residents' needs, depending upon their preferences and activities. Flexibility and control over electric lighting is essential.

Acoustic environment can be evaluated both in terms of the presence of useful or pleasing sound and the absence of undesirable sound. In the short term, unwanted noise can cause activity disturbance, speech interference and disturb rest, relaxation and sleep. In the longer term there is evidence of more insidious health effects, because the presence of noise can cause increased levels of stress hormones, increasing the risk of cardiovascular effects (heart disease and hypertension). The distress caused by neighbour noise issues and poor sound insulation between homes is well documented, although noise complaint statistics generally underestimate the extent of dissatisfaction. According to the UK National Noise Attitude Survey (2000), 81% of respondents heard noise from neighbours and/or other people nearby; 37% were bothered, annoyed or disturbed to some extent; 21% of respondents reported that noise spoilt their home life to some extent, with 8%reporting that their home life was spoilt either 'quite a lot' or 'totally'. Noise is also important for the quality of sleep. Regular poor sleep can lead to serious medical conditions such as obesity, heart disease and diabetes, and has been shown to shorten life expectancy, weaken immune system, and even affect fertility. Sleep duration and quality has been linked to the risk of anxiety and depression (Ohayon, 2003), (Taylor et al., 2005), (Green Building Council, 2016). Therefore noise control in is one of the basic spatial requirements in living environments.

Controllable thermal environment is another important factor that also have direct effect on physical health. Exposure to warm conditions in excess of approximately 25 degrees is associated with increased health risks, with thermal stresses affecting the performance of our body, productivity, and mood. Similarly the incidence of heart attacks, strokes, and other respiratory diseases is directly increased by excess cold (Green Building Council, 2016).

Good air quality depends on many factors including ventilation, humidity and effects of materials. Various studies have shown the importance of air quality for learning and behaviour in schools and productivity in offices (Green Building Council, 2016). Regarding humidity, high moisture levels encourage dust mites and mould growth; this leads to an increased risk of asthma and can also increase the risk of skin conditions such as eczema (Menon, Porteous, 2011).

Aesthetic Requirements

Aesthetic requirements affect the way people feel depending on spatial aspects, furniture and objects. According to IOM (2007), in order to be healthy, people need an interesting and aesthetically pleasing environment, sensory stimulation similar to that found in the natural environment.

Kaplan and Kaplan's informational model of environmental preference, suggests that the possibility that aesthetic activities promote certain biological functions, and that human beings are healthier with them than without them (1971). Aesthetic judgements appear, indeed, to be associated with the two basic dimensions characterising the affective appraisals manifested by individuals in relation to their physical environments, the dimensions of pleasure and arousal (Galindo, Corraliza Rodríguez, 2000). Kendall argues that the way light enters a room, the colours of floors and walls, and even the shape and texture of furniture and home accessories all work together to influence how people feel and how they perform, both consciously and subconsciously.

The understanding of aesthetics is strongly based on cultural aspects. In vernacular examples aesthetics is usually connected to other functional, ergonomic aspects as a natural outcome of clever solutions, the nature and capacity of materials and the way spatial elements and objects are created. For example the aesthetic appearance of Japanese room in traditional Japanese house (Minka) is characterised by the strong relation between the room and the garden, which becomes an extension of the interior towards nature creating an aesthetic value difficult to achieve with only interior arrangements. Other than this the modular understanding of space with tatami, the partitions with ornamental transoms and decorative shelves are all spatial elements that come together in harmony creating a strong aesthetic identity.

HIERARCHY OF WELL-BEING REQUIREMENTS AND CULTURAL DIFFERENCES

The well-being framework is composed of certain requirements discussed in the previous section. As discussed earlier, well-being requirements are different from both health and survival requirements. But every group of requirement have a role on well-being related to space. On the other hand the importance of these requirements on well-being might not be the same as they all represent different dimensions of well-being. Therefore in this section it is aimed to understand which of the well-being requirements are considered essential or more important than the others and it is also intended to see the variances related to cultural differences. In this sense, the study includes a survey study conducted in two different cultural contexts. The results might be useful in order to decide on priorities regarding well-being and culture in our living environments.

Case Study: Turin and Istanbul

In order to evaluate the relationship between well-being and living environments, a questionnaire was structured by using the well-being framework proposed by Saglar Onay and Minucciani (2018). Survey studies were conducted in two different cultural contexts; Istanbul (Turkey) and Turin (Italy). In Turkey the questionnaire was applied to totally 60 people, 17 from age group 18-30, 27 from age group 30-60 and 16 from age group 60 and over. In Italy the same questionnaire was applied to totally 41 people, 10 from age group 18-30, 19 from age group 30-60 and 12 from age group 60 and over. The selection of subjects was determined in order to include different social layers from different parts of the city's population including the city centre, commercial areas, education districts and peripheral residential areas. This was seen as a decision that will increase the comprehensiveness of the research. The results obtained from every age group were proportioned as if the numbers of people in each age group were equal in order compare factual tendencies related to age.

Results

The results obtained with the survey were evaluated in two steps considering the results of two questions. Both questions were structured to understand the hierarchy between well-being requirements according to the framework of Saglar Onay and Minucciani (2018) discussed through this chapter.

The first question was intended to comprehend the direct relationship between well-being and positive conditions related to each well-being requirement. The test subjects were asked to put in order answers each representing a well-being requirement (Table 2). The answers were arranged

Table 1. Structure of the interview

Static Information	Daily Routines	Spatial Preferences/Requirements
Age, gender, work, family structure	Social life, activities	Contextual Ergonomic Functional Psychological Social Aesthetic Sensory

Table 2. First question about hierarchy of well-being requirements

Question: Well-Being in an Interior Is Mainly Related to (Put in Order of Importance):	
Answer Representing the Corresponding Requirement	**Corresponding Well-Being Requirement**
Connection to natural environment	Connection to context and nature
Having suitable spaces for all different activities	Functional requirements
Having suitable furniture for all activities	Ergonomic requirements
Having the necessary level of privacy and personal space	Psychological requirements
Having to opportunity to meet people and social encounters	Social requirements
Having the necessary level of heating, cooling, acoustic, illumination…	Sensory requirements
Having attractive and pleasing space, furniture and objects	Aesthetic requirements

with the aim of giving an all-inclusive idea of the requirement. For example for sensory requirements all the sensory input important to feel comfortable were included in the answer. Similarly the second question was arranged to understand the relationship between well-being and negative conditions related to each well-being requirement. Respondents were asked to indicate conditions that they would surely avoid in their living environments (Table 3). Just like the first questions the answers were all-inclusive as possible.

While evaluating the results, in general it was seen that the distribution of results for the context of Italy and Turkey were considerably different from each other. Evaluating the answers of the first question for all age groups only for the context of Istanbul, Turkey, it was seen that connection to context, psychological and sensory requirements were considered more important. On the other hand for Italy, the distribution of requirements did not show big differences. Each well-being requirement was equally considered important with only comparably lower values for ergonomic requirements.

Table 3. Second question about hierarchy of well-being requirements

Question: Which of Following Condition in Your Living Environment Would You Surely Avoid: (Put in Order of Importance)	
Answer Representing the Corresponding Requirement	**Corresponding Well-Being Requirement**
Visual and sensory total insulation (without any view, any connection with exterior space and so on)	Connection to context and nature
Lack of spaces available suitable for your activities in leisure time	Functional requirements
Uncomfortable furniture and accessories	Ergonomic requirements
Lack of privacy and personal space	Psychological requirements
Lack of social contacts	Social requirements
Having uncomfortable conditions of temperature, acoustics, lighting	Sensory requirements
Lack of attracting and pleasing space, furniture and objects	Aesthetic requirements

The second questions focused on the effect of lacking needs for well-being. Regarding Turkey, the hierarchy of well-being requirements related to negative conditions was significantly interesting representing a situation completely different than the results of the first question related to positive conditions (Figure 14). This time ergonomic requirements were considered to be the most important components of well-being although they were considered not to be

Figure 4. Results of the questions about hierarchy of well-being requirements related to positive conditions and negative conditions

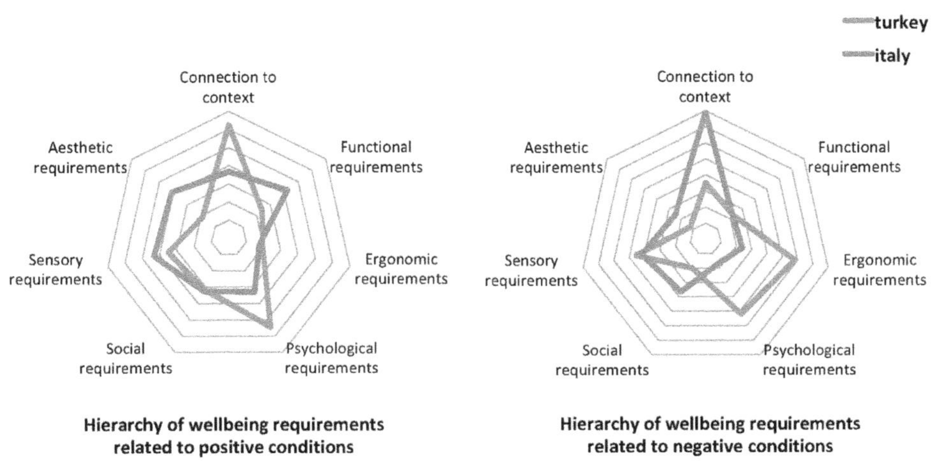

important in the first question. On the other hand psychological requirements were considered to be important both related to negative and positive conditions with a similar percentage. Similarly sensory requirements were equally important both for negative and positive conditions while connection to context has fallen to the fourth place regarding negative conditions. People were not as annoyed about the absence of connection to context as they were for the lack of uncomfortable interior components. Functional requirements followed connection to context with a similar percentage for negative and positive conditions. Social requirements followed with a slight decrease and aesthetic requirements were the last components representing again similar percentages for both negative and positive conditions.

Similarly, the results from Italy in general showed considerable difference for the second question focusing on negative conditions about well-being requirements. The lack of connection to context was considered as a very important negative effect on well-being while sensory, social and ergonomic requirements followed contextual ties.

In general regarding negative and positive conditions some requirements tended to be equally important while some reflected big differences. Although the answers for negative and positive conditions were arranged similarly (For example for ergonomic requirements the positive answer is ‘having comfortable furniture/ equipment for activities’ while the negative answer is ‘having uncomfortable furniture/ equipment for activities), respondent’s conception of hierarchy changed radically. As discussed above for Turkey the biggest difference was for ergonomic requirements and connection to context while other requirements tended to stay similar with slight differences. And for Italy the biggest difference was again for connection to context. This remarkable finding shows that well-being is not affected equally by negative and positive conditions and further research needs to focus on this aspect of well-being related to space (Figure 4).

Regarding general preferences in Italy and Turkey for both questions, the biggest differences arise for connection to context and psychological requirements. While the lack of connection to context for test subjects in Italy is considered as an important deficiency, the presence of contextual ties is considered very important in Turkey. So, the results show also big differences with the negative and positive conditions. This might arise from the fact that Italians take contextual ties for granted and does not feel a great need for it, while Turkish subjects already suffer from the lack of contextual ties and consider them important when asked positively. On the other hand psychological requirements are more important for Turkish test subjects both for negative and positive conditions as privacy is considered as a key factor in general well-being. This cultural tendency can be easily read in vernacular living environments as well as those of today.

Considering the results of the questionnaire, the most important finding in living environments is about the difference in user preferences in different cultural contexts. This proves the generally accepted statement that designers and architects cannot handle design as a global issue and in every specific context there exists some general inclinations and some various tendencies of different age groups. These differences are a result of cultural and contextual aspects. In fact this understanding was the starting point vernacular architecture, which was the outcome of a continuous accumulation of context related knowledge passing from generation to generation. This knowledge is significantly connected to the specific environment in which they are developed in order make benefit of natural sources and materials. Moreover craftsmanship is also in close relation to the potentials of these natural sources. This understanding creates a holistic approach that respond to the conditions of the environment in order to support human well-being as a natural outcome of sustainable solutions. Therefore, the simplicity and honesty of vernacular examples have a great potential in understanding the effects of design principles on well-being.

FUTURE RESEARCH DIRECTIONS

Future research needs to focus on observations and further survey studies that evaluate the effects of different spatial requirements on existing living environments. The evaluation of existing spaces might be very useful as well-being is an outcome of the short-term and long-term interaction between space and human. In this sense the results of the survey discussed in this chapter can be confronted with further studies on existing environments, which can give a broader conception of well-being.

CONCLUSION

Well-being in living environments in general needs to be considered as the focal point design activities regarding homes, schools, offices and other spaces that we spend an important portion of our lives. Therefore well-being is not an added value but it is a must to be fulfilled. It is very important to be aware of all the requirements that contribute to well-being related to space. The framework proposal, consisting of contextual, functional, psychological, social, ergonomic, sensory requirements and aesthetic requirements as basic design criteria, aims to support both theoretical and practical activities regarding well-being in all living environments, as a crucial component of sustainability. The examination of well-being through the framework also gives the possibility to analyse the effects of different requirements and their hierarchic relations in different groups and different cultures.

The fulfillment of well-being requirements sets the base for the issues to be handled in interior design. But every design problem is unique and therefore it is difficult to talk about a strict framework for design. In latest years, various building certification systems that propose design guidelines have arisen. Having therefore developed certification systems measuring and evaluating the performance of buildings in relation to their impact on the environment and its resources, we are now faced with a shift of attention on a smaller scale, linked to the performance that buildings offer not so much with respect to the environment as to the people who live in them. The well-being framework differs from these approaches with its focus on human and space interaction and its compact structure that does not scatter the issue into numerous and inconceivable small fragments but keeps it as an inter-connected whole. In this sense, the framework can be used to structure practice, education and research activities, which all together build up the interior design body of knowledge. Especially in interior design education, application of the well-being framework to design studios can make significant contributions to the quality of design proposals letting the educators and students handle the issue together with its vital components. In research activities the framework represents the intersection between architecture and humanities and therefore it can be useful to understand the effects of spatial attributes on wellbeing. The findings of research studies can be applied to practice in projects with widespread effect especially by community actions groups, social housing developers and decision makers.

REFERENCES

Afshar, P. F., Foroughan, M., Vedadhir, A., & Tabatabaei, M. G. (2016). The effects of place attachment on social well-being in older adults. *Educational Gerontology*, *43*(1), 45–51. doi:10.1080/03601277.2016.1260910

APA dictionary of psychology. (n.d.). Washington, DC: American Psychological Association.

Boyden, S. (1971). Biological determinants of optimal health. In D.J.M. Vorster (Ed.), *Proceedings of Conference The Human Biology of Environmental Change*. London: International Biology Program.

Boyden, S., & Millar, S. (1978). Human ecology and the quality of life. *Urban Ecology*, *3*(3), 263–287. doi:10.1016/0304-4009(78)90017-7

Brown, B., Perkins, D. D., & Brown, G. (2003). Place attachment in a revitalizing neighborhood: Individual and block levels of analysis. *Journal of Environmental Psychology*, *23*(3), 259–271. doi:10.1016/S0272-4944(02)00117-2

Casey, E. S. (2001). Between geography and philosophy: What does it mean to be in place world? *Annals of the Association of American Geographers*, *91*(4), 683–693. doi:10.1111/0004-5608.00266

Charytonowicz, J. (2000). Architecture and ergonomics. In *Proceedings of the Human Factors and Ergonomics Society Annual Meeting*. Sage Publishing.

Deci, E. L., & Ryan, R. M. (1985). *Intrinsic motivation and self-determination in human behavior*. Plenum. doi:10.1007/978-1-4899-2271-7

Desmet, P. M. A., & Pohlmeyer, A. E. (2013). Positive design: An introduction to design for subjective well-being. *International Journal of Design*, *7*(3), 5–19.

Freeman, K. (2017). *How wellness and well-Being went mainstream*. Retrieved from https://www.ambius.com/blog/how-wellness-and-well-being-went-mainstream/

Galindo, P. G., & Corraliza Rodríguez, J. A. (2000). Environmental aesthetics and psychological well-being: Relationships between preference judgments for urban landscapes and other relevant affective responses. *Psychology in Spain*, *4*(1), 13–27.

Green Building Council. (2016). *Health and well-being in homes*. Green Building Council.

Institute of Medicine (IOM). (2007). *Green Healthcare Institutions; Health, Environmnet and Economics*. The National Academies Press.

Katz, J., & Aspden, P. (1997). A nation of strangers? *Communications of the ACM*, *40*(12), 81–86. doi:10.1145/265563.265575

Keller H. (2016). Psychological autonomy and hierarchical relatedness as organizers of developmental pathways. *Philosophical Transactions of the Royal Society of London. Series B, Biological Sciences, 371*(1686), 20150070.

Kellert, R. (2005). *Building for life: designing and understanding the human-nature connection*. Island Press.

Kendall, D. (n.d.). *Aesthetics and Happiness: How Space Affects Well-Being*. Retrieved from www.huffpost.com

Keyes, C. L. M. (1998). Social Well-being. *Social Psychology Quarterly, 61*(2), 121–140. doi:10.2307/2787065

Keyes, C. L. M., & Shapiro, A. D. (2004). Social well-being in the United States: A descriptive epidemiology. In O. G. Brim, C. Ryff, & R. Kessler (Eds.), *How healthy are we? A national study of well-being at midlife* (pp. 350–372). University Chicago Press.

Kong, F., Xue, S., & Wang, Y. (2016). Amplitude of low frequency fluctuations during resting state predicts social well-being. *Biological Psychology, 118*, 161–168. doi:10.1016/j.biopsycho.2016.05.012 PMID:27263835

Koningsveld, E. A. P., Settels, P. J. M., & Pikaar, R. N. (2007). *Meeting diversity in ergonomics*. Elsevier. doi:10.1016/B978-008045373-6/50001-5

Kraut, R., Patterson, M., Lundmark, V., Kiesler, S., Mukhopadhyay, T., & Scherlis, W. (1998). Internet paradox: A social technology that reduces social involvement and psychological well-being? *The American Psychologist, 53*(9), 1017–1031. doi:10.1037/0003-066X.53.9.1017 PMID:9841579

Kupfer, J. (1987). Privacy, autonomy, and self-concept. *American Philosophical Quarterly, 24*(1), 81–89.

Lyubomirsky, S., Sheldon, K. M., & Schkade, D. (2005). Pursuing happiness: The architecture of sustainable change. *Review of General Psychology, 9*(2), 111–131. doi:10.1037/1089-2680.9.2.111

Menon, R., & Porteous, C. (2011). *Design guide: Healthy low energy home laundering*. Mackintosh Environmental Architecture Research Unit.

Naci, H., John, P. A., & Ioannidis, M. D. (2015). Evaluation of wellness determinants and interventions by citizen scientists. *Journal of the American Medical Association, 314*(2), 121–122. doi:10.1001/jama.2015.6160 PMID:26068643

Namazian, A., & Mehdipour, A. (2013). Psychological demands of the built environment, privacy, personal space and territory in architecture. *International Journal of Psychology and Behavioral Sciences*, *3*(4), 109–113.

Nie, N., Hillygus, S., & Erbring, L. (2002). Internet use, interpersonal relations and sociability: findings from a detailed time diary study. In B. Wellman & C. Haythornthwaite (Eds.), *The Internet in Everyday Life*. Blackwell Publishing. doi:10.1002/9780470774298.ch7

Ohayon. (2003). The effects of breathing-related sleep disorders on mood disturbances in the general population. *J Clin Psychiatry, 64*(10), 1195-200.

Pallasma, J. (2005). *The Eyes of the Skin*. John Wiley &Sons.

Pam, M. S. (2013). Security. In *PsychologyDictionary.org*. Retrieved from https://psychologydictionary.org/security/

Petermans, A., & Pohlmeyer, A. E. (2014). Design for subjective well-being in interior architecture. *Proceedings of the 6th Annual Architectural Research Symposium*, 206–218.

Phesant, S. (2003). *Bodyspace: Anthropometry, ergonomics and the design of work*. Taylor & Francis.

Poe, M. R., Donatuto, J., & Satterfield, T. (2016). Sense of place: Human well-being considerations for ecological restoration in Puget Sound. *Coastal Management*, *44*(5), 409–426. doi:10.1080/08920753.2016.1208037

Rollero, C., & De Piccoli, N. (2010). Does place attachment affect social well-being? Revue Européenne de Psychologie Appliquée/European. *Review of Applied Psychology*, *60*(4), 233–238. doi:10.1016/j.erap.2010.05.001

Ryan, R. M., & Deci, E. L. (2000). Self-determination theory and the facilitation of intrinsic motivation, social development, and well-being. *The American Psychologist*, *55*(1), 68–78. doi:10.1037/0003-066X.55.1.68 PMID:11392867

Saglar Onay, N. (2017). Analysing Spatial Potentials Of Historic Buildings. Santarcangelo di Romagna, Rimini: Maggioli Editore.

Saglar Onay, N., Garip, B. & Garip, E. (2017). A flexible user centered design model for social housing units. *Tafter Journal, 97.*

Saglar Onay, N., Garip, B., & Garip, E. (2019). *Re-coding homes through flexible interiors: Emerging research and opportunities*. IGI Global.

Saglar Onay, N., & Minucciani, V. (2018). Well-being framework as a contributor to sustainability. *WIT Transactions on Ecology and the Environment, 217*, 699–710. doi:10.2495/SDP180591

Salvendy, G. (2012). *Handbook of human factors and ergonomics*. John Wiley & Sons. doi:10.1002/9781118131350

Schneier, B. (2008). The psychology of security. In S. Vaudenay (Ed.), *Africacrypt, LNCS 5023* (pp. 50–79). Springer-Verlag.

Schulz, C. N. (1980). *Genius Loci: Towards a Phenomenology of Architecture*. Academy Editions.

Shah, H. & Peck, J. (2005, July). Well-being and the environment. *Nef.*

Shapiro. (1998). *Control therapy*. Wiley.

Shklovski, I., Kiesler, S., & Kraut, R. (2006). *The Internet and social interaction*. Carnegie Mellon University.

Taylor, D. J., Lichstein, K. L., Durrence, H. H., Reidel, B. W., & Bush, A. J. (2005). Epidemiology of insomnia, depression and anxiety. *Sleep*, *28*(11), 1457–6464. doi:10.1093leep/28.11.1457 PMID:16335332

Taylor, S. E., & Brown, J. D. (1998). Illusions and well being: A social psychological perspective on mental health. *Psychological Bulletin*, *103*(2), 193–210. doi:10.1037/0033-2909.103.2.193 PMID:3283814

The UK National Noise Attitude Survey 1999/2000. (2002). *Noise Forum Conference*.

Triandis, H. C. (1995). *Individualism and collectivism*. Westview.

Unwin, S. (2009). *Analyzing Architecture*. Routledge. doi:10.4324/9780203880906

Wilson, J. R. (2000). Fundamentals of ergonomics in theory and practice. *Applied Ergonomics*, *31*(6), 557–567. doi:10.1016/S0003-6870(00)00034-X PMID:11132041

Yiengprugsawan, V., Banwell, C., Takeda, W., Dixon, J., Seubsman, S., & Sleigh, A. C. (2015). Health, happiness and eating together: What can a large Thai cohort study tell us? *Global Journal of Health Science*, *7*(4), 270–277. doi:10.5539/gjhs.v7n4p270 PMID:25946941

Zumthor, P. (2006). *Atmospheres*. Birkhauser.

KEY TERMS AND DEFINITIONS

Framework: A complete structure of rules and principles for a certain task.

Chapter 6
Well-Being and the Effect of Age in Interiors

ABSTRACT

This chapter starts from the assumption that understanding of well-being evolves over time (that's during history) and with time (that's in the lifecycle of people). These differences have not been considered with necessary attention until now, and a study carried by the authors is presented here: several interviews performed in different contexts and with differently aged people have shown some differences between young people, adults, and the elderly. In particular, results show that psychological needs change, overcoming functional or aesthetics requirements. Among young people, connection to context is not so important, as their social life is performed in social media more than in real, physical environments. Hierarchies evolve too, showing interesting results. At the same time, cultural context shows several, unexpected differences (e.g., about relation with natural and built context). These results invite one to design interiors as flexible spaces, available to be adapted over time and with time.

INTRODUCTION

When do designers deal with elderly? They do only when they are asked to design hospices, or retirement homes and when elderly become disabled and their health circumstances decrease. And when do designers deal with young people's needs? They do only in designing schools (then referring

DOI: 10.4018/978-1-7998-4231-6.ch006

to children and very young people), or when youngest have psychological sicknesses or physical injuries.

According to Hillman (1999), although aging is a given of the human condition, it is subject to several biases. He states that, while biology considers aging as a process leading to uselessness, the third age should be also considered as a structure, with specific features. Old people are in balance between lasting and leaving. Their desires and their perspectives aren't the same as young or middle age people. Logically their idea of happiness, as well as their feeling of well-being and their concept of life should differ from other sectors of the population. Similarly, the needs of young people are also subject to preconceptions. It is taken for granted that they love to socialize and engage in physical activity, and that their well-being degree is naturally high when they do not have health problems. (Minucciani, Saglar Onay, 2019).

A considerable number of researches argue that all individuals in society have different but stable levels of well-being, which are not supposed to change across the lifespan. According to Ulloa, Moller and Sousa-Poza (2013), changes in the environment, although important for short-term well-being, lose salience over time through processes of adaptation, and have small effects on long-term subjective well-being. This process, called hedonic adaptation, is defined as "the process by which individuals return to baseline levels of happiness following a change in life circumstances" (Lucas, 2007). This view generally routes from the set point theory, which argues that individuals are born with a predisposition to a certain level of happiness, based on genetics and personality. On the other hand, according to Lyubomirsky, Sheldon, and Schkade (2005), other than the happiness set point, life circumstances and positive cognitive, behavioural, and goal-based activities play an important role on subjective well-being which might account for a significant portion, 50 percent of individual differences. Moreover, life circumstances and our activities are highly affected by living environments (Minucciani, Saglar Onay, 2019).

Ulloa, Moller and Sousa-Poza (2013), describe the mainstream theories on the relation between aging and well-being from the perspectives of economics, psychology and gerontology. Among these, insights from psychology and gerontology are important in terms of space and aging. According to Ulloa and others, even life-changing events, such as marriage or the death of a loved one, only have temporary effects on the individual's life satisfaction, which after a while will return to the original, biologically determined level.

According to this perspective, even after major life events like winning the lottery or becoming paraplegic, individuals return to a 'baseline' level of well-being (Brickman et. al., 1978; Kahneman, 1999). This theoretical framework was then complemented by Costa and McCrae's (1980) personality theory of well-being, which specified that individuals are born with certain personality traits that do not change significantly across the lifespan. Lucas (2007), however, found some contradictory evidence; namely, that well-being levels do in fact change over time and these changes appear to be permanent. The major events that precipitate such changes include marriage, unemployment and disability (Lucas, 2007).

AGE AND WELL-BEING

Common sense dictates that well-being should decrease among older individuals, not least because health diminishes substantially. That is, advancing age compromises not only physical but also mental capabilities. However, according to much gerontological research, "well-being seems to be unaffected by the adverse contexts brought on by the aging process" (Mroczek, Kolarz, 1998). This phenomenon, known as the satisfaction paradox or "stability despite loss" paradox, tries to explain the disconnect inherent in the relatively high levels of subjective life satisfaction reported by elderly living in objectively bad conditions. The gerontological literature notes that this so-called paradox might be caused by age-cohort effects – that is, older people may report higher levels of life satisfaction because of the lower expectations of a particular generation (Walker,2005).

Other researchers of the elderly have reached to similar conclusions. For instance, Argyle (2001) found evidence that happiness increases slightly with age, mainly due to a declining goal-achievement gap. In other words, as time goes by, individuals realize that their expectations were probably set too high in their younger years and learn to accept the reality of their lives. Likewise, Diener et al. (1999, p. 291) concluded that "life satisfaction often increases, or at least does not drop, with age".

Authors in different fields have empirically analysed the relation between age and well-being and three main trends have been identified: a U-shape (convex), an inverted U-shape (concave) and linearity (Ulloa, Moller and Sousa-Poza, 2013). In recent years, a number of authors, especially economists, have identified a U-shaped relation between age and reported life satisfaction. In these analyses, well-being is believed to reach its minimum between a person's mid-30s and early 50s (Van Landeghem, 2012; Blanchflower, Oswald, 2008). Contrary to the U-shape, very limited evidence on an inverted U-shape exists. A linear relation between reported well-being and age has also been observed, though it can be constant throughout the lifecycle or sloping upward or downward. However one could thus conclude that the age effect on subjective well-being is rather small. It should, however, be noted that the difference in well-being between the top and bottom of the curve is about one fifth to one third of the standard deviation of life satisfaction scores. (Ulloa, Moller, Sousa-Poza,2013).

AGE WELL-BEING AND THE BUILT ENVIRONMENT

Evaluating the coverage and findings of previous research about well-being and age in our living environments, it clearly emerges that the role of architecture and interior design is underestimated. The studies that focus on the relation between age and well-being lack the environmental and spatial dimension. On the other hand the studies that consider well-being related to space and age focus on only specific groups like the elderly or teenagers with problems. In this sense there is serious need for research that focus on well-being requirements related to space, age and culture (Minucciani, Saglar Onay, 2019).

According to previous studies (Saglar Onay & Minucciani, 2018), the physical features of a space can affect well-being in relation with contextual, functional, social, psychological, sensory, ergonomic and aesthetic requirements that are in turn affected by the cultural context. Although tendency to positively consider one's own quality of life is genetically affected, well-being requirements change over time according to our mental and physical conditions.

Normative developmental and life course transitions associated with growing old may be accompanied by concomitant changes in the use and meaning of space (Bronfenbrenner, 1979, 1999; Oswald & Wahl, 2003). Framed in terms of an ongoing person-context dynamic, the transaction between the person and the social/physical environment characteristically becomes increasingly multifaceted from childhood through adult life and into old age as a result of the accumulation of layer upon layer of life experiences and place meanings. Interior spaces are especially important during certain phases of this lifelong dynamic (Rowles, Oswald, Hunter, 2004). For instance, home space is generally the fulcrum of life space during childhood when the child's world may be limited to the dwelling and its immediate surroundings. With the transition into adolescence, the challenge for each developing child is to find a balance between basic needs for a secure, safe, and stable base (i.e. the interior living environment of home) on the one hand and higher-order needs encouraging exploration, stimulation, and environmental mastery on the other (Balint, 1955).

Early Life

The origins of modern psychology are rooted in the notion that stressful early-life experiences and events that exceed a child's ability to cope (Cohen, Tyrrell, & Smith, 1991) negatively impact mental health. Moreover, emerging work in both the behavioral and biological sciences has shown that early-life stress also negatively impacts physical well-being (Fagundes, Glaser, & Kiecolt-Glaser, 2013; Miller, Chen, & Parker, 2011; Perroud et al., 2011). When linking early-life stress to inflammation, psychologists have focused on childhood stressors that are both severe and chronic, because those are the type of stressors that are most likely to have a lasting impact on the immune system (Segerstrom & Miller, 2004; Fagundes, Way, 2014). Other than mental and physical conditions the effects of economic factors have also been discussed. The accumulation of risk model suggests that the cumulative effect of living in a low income family throughout childhood affects subjective well-being (defined as 'a person's cognitive and affective evaluations of his or her life' by Diener, Lucas, & Oshi, 2002) in adolescents (Kuh D, Ben-Shlomo Y, Lynch J, Hallqvist J, 2003).

Evidence from neuroscience and developmental psychology indicates that the first years of life are crucial for brain development and may be a time when children are particularly sensitive to their environment (Lipina, Colombo, 2009) But research on how the environment should respond to the needs of young children is limited and the existing research is mostly based on school environments and the relationship between learning and the environment. The school environment, as a context of learning, has been found to play an important role in children's social, emotional and behavioural well-being (Gutman and Feinstein, 2008). In several countries, a focus on student or learner well-being, mental health, and/or social and emotional learning in educational settings is becoming increasingly systematic, with various frameworks, initiatives and programmes being developed and implemented (e.g. Australian Government Department of Health and Ageing 2009;

Australian Government Department of Health, Ageing 2010; Bernard 2004; Department for Education and Skills 2007; Durlak et al. 2011; Gammage 2008; Mashford-Scott, Church, Tayler, 2012). It is known that young children benefit intellectually from experiencing their environments. Visual, tactile, and auditory senses are crucial in learning and need to be considered in the design of the built environment (Kaup, Kim, Dudek, 2013).

Considering specific features of the early life, we should note that the well-being needs of young children are not sufficiently discussed. Because their needs are not limited to learning environments as they have great tendency to learn wherever they go. Also home environments and other environments that they live in play an important role on their development and happiness. Therefore the concept of well-being need to be handled together with all its dimensions also for young children.

Adolescence

Adolescents who report greater subjective well-being (SWB) are better at schools, have better relationships, and have less mental health problems (Suldo, Thalji, Ferron, 2011; Keyes, 2006), and these benefits carry into adulthood (De Neve, Oswald, 2012; Kansky, Allen, Diener, 2016). Recent qualitative research found that adolescents endorsed SWB as involving multiple aspects of life, such as having good relationships and having life goals (Navarro, Montserrat, Malo, Gonzalez, Casas, 2017).

To examine the potential role of subjective well-being as a long-term predictor of critical life outcomes, Kansky, Allen and Diener (2016) examined indicators of positive and negative affect at age 14 as predictors of relationship, adjustment, self-worth, and career outcomes a decade later at ages 23 to 25, controlling for family income and gender. Results suggest that adolescent positive affect may have a strong correlational and predictive relationship with many important life outcomes, while negative affect is less strongly associated with parallel outcomes. Positive affect specifically may be predicting desirable relative changes in critical social, adjustment, self-worth, and career out-comes over time, consistent with Fredrickson's (1998, 2001) broaden-and-build model (Kansky, Allen and Diener, 2016)

The Good Childhood Inquiry (Layard and Dunn, 2009) surveyed around 8,000 young people (14-16 years) across 16 areas in England and within school or educational establishments, asking them what they thought were the most important things that make for a good life for young people, and what in their view prevented young people from having a good life. An analysis of their replies revealed three overarching themes with respect to how young people perceived a 'good childhood': the quality of relationships they had with others, safety, and freedom. Aside from these themes there were 10 components identified as important for a 'good life': family; friends; leisure opportunities; school, education and learning; their own behaviour; their local (physical) environment; their local community; money; their own attitudes; and health, particularly mental and emotional aspects of health. So according to the results of this significant survey, it can be assumed that the environment has an important role on adolescent well-being especially regarding its social and psychological dimensions. In this sense interiors that are often used by young people need to create opportunities for social interaction and set the balance between private and public space, which can play an important role in social encounters. The components for the good life defined by the young people themselves also shows that they are aware of their role in their own well-being and this necessitates the conception of freedom also for the environment they live in. In other words the environment should give them the possibility to decide for themselves, to be creative and to get involved in activities that they desire. Because at this period young people start to be more selective and want to demonstrate their ideas, decisions and tastes more than before and this plays an important role in expressing themselves in a proper way and defining their identities.

Hendry and Kloep (2007), argues that the uncertainty, increased responsibility, and demands on self-direction that typify emerging adulthood can contribute to declines in subjective well-being and mental health. However, emerging adulthood has conversely been described as a window of opportunity for positive change (Arnett 2007), and some studies have shown that average levels of mental health and subjective well-being tend to improve in emerging adulthood (Galambos et al. 2006). So, it can be assumed that well-being in adolescence is dependent on factors that are not necessary similar to those of other age groups because adolescence is the passage to the period of independence and construction of the upcoming life with all its dimensions. In this sense the presence of factors that contribute the positive effect and the absence of factors that lead to negative effect can influence both the levels of well-being in adolescence and periods that follow this critic stage.

Adulthood

Regarding adult well-being, many research focus on methods for measuring general well-being taking into account different approaches. According to the research conducted with 36 adults in the UK by Al-Janabi, Flynn and Coast (2012), measures of well-being were defined in terms of the capability approach which advocates assessing capability (what an individual can do) rather than functioning (what they actually do) to avoid imposing a particular idea of what a good life constitutes and to reflect the importance of freedom to choose (Sen, 1993; Alkire, 2012). Five attributes of capability well-being were identified as: “stability”, “attachment”, “achievement”, “autonomy” and “enjoyment”. According to this research adult well-being is a consequence of people’s capability to obtain a sense of continuity in life (in terms of friends, work and location); the capability to reach to the feelings of love, support and social contact, the capability to be independent, the capability to both move forward in their life and attain their goals and the capability to enjoy life by different factors such as the presence of families, friends, pets, leisure activities and the countryside in the informants’ lives. These measures represent a blend of the two views for well-being that are the hedonic and eudaimonic traditions.

Other research on adult well-being focus on the existence of positive and negative effects on well-being following the understanding that well-being is experiencing a high level of positive affect, a low level of negative affect, and a high degree of satisfaction with one’s life (Deci and Ryan, 2008). In adulthood one of the most important factors that affect well-being is mental disorder, which is mostly shown to decline by age. The Human Rights and Equal Opportunity Commission (1993, pp. 908, 925) concluded that people with mental illness are among the most vulnerable and disadvantaged in community; they may experience stigma and discrimination in many aspects of their lives. Mental illness can be transient; some people experience their illness only once and fully recover. For others, it recurs throughout their lives. According to The 1997 National Survey of Mental Health and Well-being of Adults (SMHWB) in Australia, the prevalence of mental disorder generally decreased with age. Young adults aged 18–24 years had the highest prevalence of mental disorder (27%), declining steadily to 6.1% of those aged 65 years and over. The highest rate of anxiety disorders (16%) was observed among women aged 45–54 years. For men, the prevalence of anxiety disorders varied little with age until age 55, after which the prevalence declined. People who are depressed lose their enjoyment of life, lack energy and concentration, and may suffer sleep and appetite disturbances (McLennan, 1997). Regarding the positive effect on well-being there are research that focus on the importance of adult learning (Cozolino, Sprokay, 2006; Field, 2009), research that consider the importance of interventions like mindfulness that have appositive effect on well-being (Davis, Morris, Drake, 2016) and other researches that focus on certain aspects of our life that could effect the conception of well-being. On the other hand research focusing on only adult population do not consider the environmental dimension of well-being. Moreover the age range in different research show considerable differences as it is difficult to draw strict lines between age groups.

Later Life

Compared with other age groups, well-being research dedicated to the elderly is more comprehensive. This maybe related to the common sense that elderly has more health problems than the other age groups in the society. On the other hand, well-being for different age groups is not simply synonymous of "health", but it maintains all its complexity – as well as design for elderly is not simply a "barrier-free" or "inclusive" design, although these elements are very important. Several scholars (Dehan, 2007; Regnier, 2002; Sloane et al., 2002) have focused on psychological, sensory and social well-being requirements as crucial for elderly.

Although more and more research focus on well-being in later life, it is commonly found that the well-being of older adults worldwide is very high, in spite of declining health and other losses (Myers & Diener, 1995). High well-being is found even in older people with poor living conditions (Baltes, Baltes, 1990) or acute ill-health (Pearlman & Uhlmann, 1991). Given that, with ageing, there are declines in areas such as health and social interaction (Pearlin, 1994), and that health and social interaction are two of the most important things to older people (Cross & Marhs, 1991; Keller, Leventhal, & Larson, 1989; Ryff, 1989). Wallace (1992) showed that older people are more likely to accept that some goals are no longer attainable than to be depressed about it. Acceptance of one's life as it has been lived is the key element in what Erikson (1959/1980) designated as the route to achieving integrity, the goal of the last stage of life. According to Erikson's theory of psychosocial development, if people do not accept the way they have lived their lives, they may give way to despair. Conversely, a high level of acceptance of life would be associated with high well-being (Ranzijn, R., Luszcz M, 1999). On the other hand some research focus on the disadvantages of later life. Involuntary environmental changes (either the need for reconstruction of an existing dwelling or forced relocation to a new residence) can result in increased mortality rates, seriously compromise functional health, reduce life satisfaction, and undermine the psychological well-being of elders, especially if the elders are already vulnerable in terms of declining health or financial problems (Danermark, Ekstrom,1990; Lawton, Yaffe, 1970; Pruchno, Resch, 1988).

Among well-being research for later life we come across several studies that focus on different dimensions of well-being related to the living environments. Many of these studies focus on home environments as elderly tend to spend most of their time at home. As old age approaches, the spatial extent of the physically inhabited environment tends to become more restricted. This represents far more than a developmental regression to the limited realm of space inhabited in childhood. The gestalt of the use and meaning of space may become qualitatively different (Rowles, 1978). Considering the dynamics of elders' everyday life, there is evidence to support a progressive reduction of the spatial range of activities, especially in very old age (Moss &Lawton, 1982; Rowles, 1978). As a result, elders tend to spend more time at home in interior environments than do younger people. Recent data show that elders (65 years of age and older) in Germany spend, on average, 80% of each day at home (Küster, 1998). Other than the home environment, the character of the neighbourhood has an important impact on the mobility and quality of life of the elderly. Several studies have been carried out on specific well-being requirements of older people, and some of them are specifically related to physical elements and to space that they live in (Saglar Onay, Minucciani, 2018).

Based on Nazroo et al. (2005) framework about the main factors influencing quality of life in later life, McCormick (2009) identified five essential elements for older people well-being: Resilience, Independence, Health, Income and wealth, Having a role and having time. Among them, Independence is strictly related to physical environment, being housing and home particularly important for people over 75 who spend a great part of their time in domestic interiors. Also the role of environment is connected to the Independence issue, especially for elderly living alone and often lacking activity: too busy areas undermine well-being due to the traffic, the noise, the lack of green spaces and of opportunity of social contacts (Allen 2008), as well as isolation impacts on feeling of safety and on willingness to go out.

Burton et al. (2011) investigated the environment impact on healthy aging, developing a "Neighbourhood Design Characteristics Checklist" to support studies and design. They identified three general categories of place-related well-being constructs (functional, social and emotional): according to their research and surveys, well-being in later life seems to be positively related to amount of greenery, to block size and to setback of dwellings from the street, negatively related to density, and depending on location (e.g. villages, large towns), street pattern, street topography.

Several scholars point at psychological needs in later life, underlining that for older people self-identity is strictly related to place identity (Peace et al., 2005; Smith, 2009), while Jonathan Zaff and Ann Sloan Devlin (1988) point at the "sense of community", noting as few researchers have considered it beyond a mental health setting, or have studied how environmental aspects can promote or discourage it. They focused on the fact that physical environments can make behaviour within the community less or more controllable: and elderly people, in general, "seem to have a strong sense of residential territoriality and distinguish between communal and individual spaces" (Hartman et al., 1975). A research led by Zaff and Sloan Devlin, (1998) among elderly people in order to compare the responses of high-rise and garden apartment residents, has shown that garden apartments promote a greater sense of community than high-rises, and in particular that a sense of belonging is more stronger where there is considerable semi-public space, that helps develop informal social networks (Newman, 1973). Furthermore, the proximity of apartments and lack of defensible space could lead to conflicts, thus clear territorial demarcations such as courts helps to regulate behaviours and to support a sense of community.

There are also research studies that focus on the environment in general, not only homes and surroundings. These studies mostly focus on the differences in spatial experience related to the physical restrictions. Lawton (1974) has concentrated on how the built environment can restrict options for older people. He uses the more neutral term 'environmental pressure' to describe the impact that poorly designed homes, public buildings and places can have on people's lives. He defines environmental pressure as a state induced by the necessity of dealing with environments built for younger, fitter people which may render the older person more vulnerable, or more docile and accepting of environmental constraints (Lawton, 1974, Hanson,2004).

Andersson (2011), based on the evidence that architecture can create supportive environments for older people and it has a therapeutic effect, explores the appropriate space for aging and his conclusions underline the role of the aesthetic dimension of architectural space; the importance of a location within existing, urban built environments and the positive effects of the close proximity to nature and greenery.

CASE STUDY

In order to evaluate the relationship between age, well-being and living environments, a questionnaire was structured by using the well-being framework proposed by Saglar Onay and Minucciani (2018). The framework, which tries to handle the issue of well-being with a methodological approach, studies the general concept of well-being in interior space in terms of connection to context and functional, ergonomic, psychological, social, sensory requirements. After recent field research, the framework was extended to include also aesthetic requirements as they were evaluated to have an important effect on general well-being related to physical environment (Figure 1). These requirements and the framework in general differentiate from other studies especially with its focus on interiors and human needs. These requirements are also different than general requirements of design for well-being as they are strongly connected to space and context. Therefore, the well-being framework for interiors aims to define the spatial dimensions of well-being in order to evaluate the importance of every spatial aspect that has an impact on our well-being related to interior space. On the other hand the framework does not attempt to isolate these spatial aspects or requirements from each other as they are always in relation to each other and they are always subject to change in relation to the dynamics of everyday life. Moreover all these requirements are subject to great cultural differences regarding users with different backgrounds.

The well-being framework for interiors can be used both as an evaluation method for existing environments and a roadmap for designing new living environments that can promote well-being. The structure of the framework is based on contextual, functional, psychological, social, sensory, aesthetic and ergonomic requirements, which are always in relation to each other. Connection to context has an important role as the factor that makes every interior unique with its location and relationship to the rest of the world. Functional requirements are concerned with creating spatial solutions that support human activities. Psychological requirements are more indirect as they deal with the outcomes of the interaction between human and space. Social requirements deal with the spatial opportunities for the level of

Figure 1. Well-being Framework for Interiors
(Saglar Onay, Minucciani, 2018).

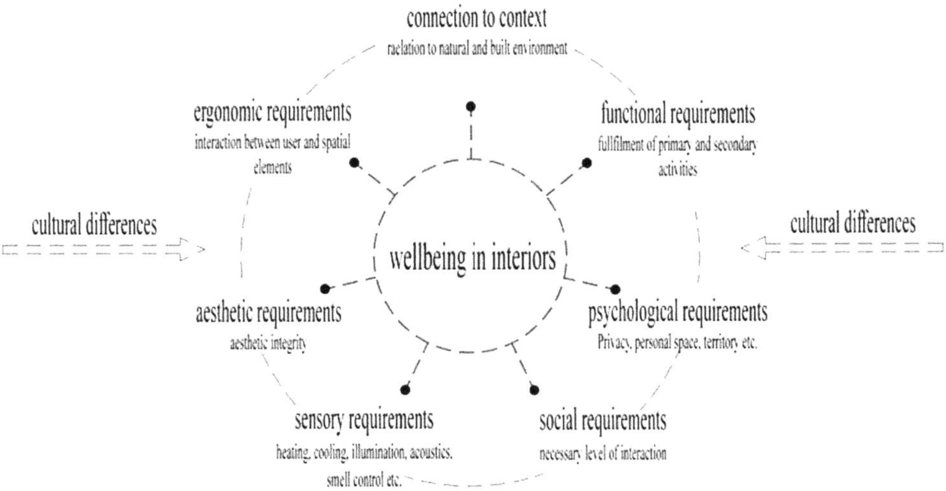

interaction between people. Sensory requirements are concerned with both quantitative factors like heating, illumination, acoustics, sound control etc. and their reflections on human senses. Aesthetic requirements are also important as they refer to the integrity and harmony of all spatial elements that can be pleasing to human senses and can promote well-being. Ergonomic requirements are more concrete and they refer to the relationship between all interior components and human body and capabilities.

The questionnaire was structured to collect data about personal aspects, daily routines and preferences about well-being requirements (Table 1). Survey studies were conducted in two different cultural contexts; Istanbul (Turkey) and Turin (Italy). In Turkey the questionnaire was applied to totally 60 people, 17 from age group 18-30, 27 from age group 30-60 and 16 from age group 60 and over. In Italy the same questionnaire was applied to totally 41 people, 10 from age group 18-30, 19 from age group 30-60 and 12 from age group 60 and over. In both countries, the selection of subjects was

Table 1. Structure of the interview

Static Information	Daily Routines	Spatial Preferences/Requirements
Age, gender, work, family structure	Social life, activities	Contextual Ergonomic Functional Psychological Social Aesthetic Sensory

determined in order to include different social layers from different parts of the city's population including the city centre, commercial areas, education districts and peripheral residential areas. This was regarded as a factor that will increase the comprehensiveness of the research. The results obtained from every age group were proportioned as if the numbers of people in each age group were equal in order compare factual tendencies related to age.

Results

The results obtained with the survey were evaluated in two steps considering the results of two questions. Both questions were structured to understand the hierarchy between well-being requirements according to the framework of Saglar Onay and Minucciani (2018). The first question was intended to comprehend which well-being requirements were more important in our living environments. The test subjects were asked to put in order answers each representing a well-being requirement (Table 2). The answers were arranged with the aim of giving an all-inclusive idea of the requirement. For example for sensory requirements all the sensory input important to feel comfortable were included in the answer. Similarly the second question was arranged to understand the relationship between well-being and negative conditions related to each well-being requirement. The answers were structured to represent

Table 2. First question about hierarchy of well-being requirements

Question: Well-Being in an Interior Is Mainly Related to (Put in Order of Importance):	
Answer Representing the Corresponding Requirement	**Corresponding Well-Being Requirement**
Connection to surrounding environment	Connection to context and nature
Having suitable spaces for all different activities	Functional requirements
Having appropriate furniture for all activities	Ergonomic requirements
Having the necessary level of privacy and personal space	Psychological requirements
Having the opportunity to meet people and social encounters	Social requirements
Having the necessary level of heating, cooling, acoustic, illumination…	Sensory requirements
Having attractive and pleasing space, furniture and objects	Aesthetic requirements

negative conditions about each well-being requirement (Table 3). For both questions the results were grouped in order to see the variances between different age groups, which were classified as 18-30, 30-60, and 60 over representing the young, middle age and elderly populations.

First Question: Evaluation of Positive Conditions About Well-Being Requirements

Test subjects were asked to choose the most important 4 requirements (represented by the appropriate answers) and put them in order of importance. While the first preference was given 4 points, the last was given 1 point. The graphs were constructed to show the sum of total points for each requirement by distinguishing the preferences of subjects from different age groups. The values on the graph show the first three preferences for each age group (represented by different colours) for both Turkey and Italy.

While evaluating the results, in general it was seen that the distribution of results for the context of Italy and Turkey were considerably different from each other. Evaluating the answers of the first question for all age groups only for the context of Istanbul, Turkey, it was seen that connection to context, psychological and sensory requirements were considered more important. While evaluating the variances for different age groups in Istanbul, it was noted that the preferences for functional and ergonomic requirements decreased with age while those for sensory requirements and the will to be connected to context increased slightly with age. However the age effect on well-being

Figure 2. Results of the first question for different age groups (both for Italy and Turkey) about hierarchy of well-being requirements related to positive conditions (for each age group the most important 3 requirements are indicated orderly with numbers)

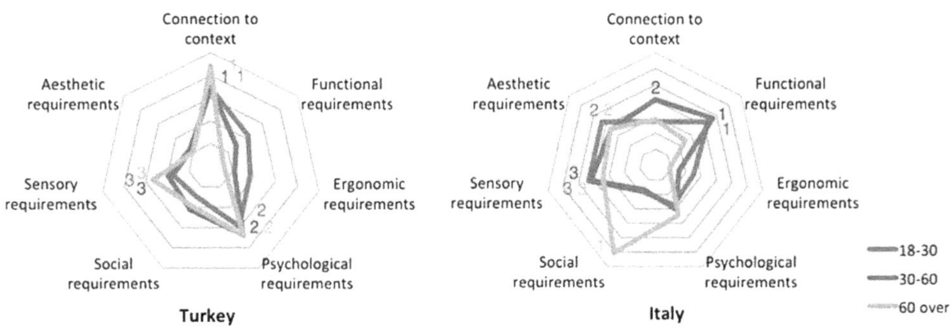

requirements can be considered rather small. The results obtained from the context of Turin, Italy were notably different from the ones obtained from Istanbul. The distribution of preferences for different requirements did not show big variances. Each well-being requirement was equally considered important with only comparably lower values for ergonomic requirements. Evaluating the effect of age in the context of Italy, biggest differences were noted for social and functional requirements while other requirements did not show big variances. Preferences for social requirements notably increased for elderly whereas those for functional requirements decreased (Figure 2).

Second Question: Evaluation of Negative Conditions About Well-Being Requirements

The second questions focused on the importance of lacking needs for well-being. The graphs for the second question were constructed in the same way as the first question. The test subjects were asked to choose the four conditions each related to a well-being requirement (as shown in Table 3) and put them in order of importance. The first condition that they would surely avoid was

Table 3. Second question about hierarchy of well-being requirements

Question: Which of Following Condition in Your Living Environment Would You Surely Avoid: (Put in Order of Importance)	
Answer Representing the Corresponding Requirement	**Corresponding Well-Being Requirement**
Visual and sensory total insulation (without any view, any connection with exterior space and so on)	Connection to context and nature
Lack of spaces available suitable for your activities in leisure time	Functional requirements
Uncomfortable furniture and accessories	Ergonomic requirements
Lack of privacy and personal space	Psychological requirements
Lack of social contacts	Social requirements
Having uncomfortable conditions of temperature, acoustics, lighting	Sensory requirements
Lack of attracting and pleasing space, furniture and objects	Aesthetic requirements

given 4 points while the last was given 1 point. The graphs were constructed to show the sum of total points for each requirement by distinguishing the preferences of subjects from different age groups. The values on the graphs (Figure 3) orderly show the negative conditions that the subjects would avoid for each age group (represented by different colours) for both Turkey and Italy.

Regarding negative conditions, psychological and ergonomic requirements were considered very important for the context of Turkey. The effect of connection to context was not as high as the first question. Different age groups did not show considerable differences for the context of Turkey. On the other hand, the results from Italy in general showed considerable difference for the second question focusing on negative conditions about well-being requirements. The lack of connection to context was considered as a very

Figure 3. Results of the second question for different age groups (both for Italy and Turkey) about hierarchy of well-being requirements related to negative conditions (for each age group the most important 3 requirements are indicated orderly with numbers)

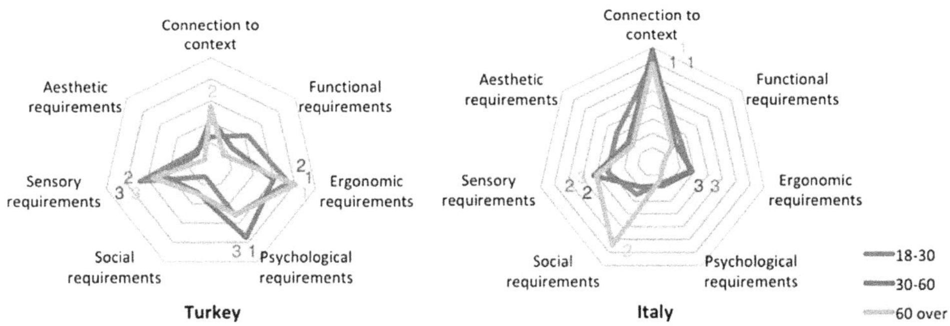

important negative effect on well-being. Regarding the age effect in Italy for the second question, the biggest difference was again for social requirements. The lack of social contacts was considered as the most important problem by the elderly people. Just like the results of the first question, the results obtained from Italy and Turkey showed big differences (Figure3).

Observing the results obtained from the case study, some of the well-being requirements change with age while some of them do not show big differences. For example for Italy social requirements considerably increase with age. On the other hand, for Turkey sensory requirements and connection to context become more important with increasing age. Therefore the comparison between different age groups rises as an important issue because of the different results obtained through the interview. However it is seen that the effect of culture on well-being requirements is much more important than the effect of age. Both general tendencies and those related to age change significantly for the two different cultural contexts. Regarding general preferences in Italy and Turkey for both questions, the biggest differences arise for connection to context and psychological requirements. While the lack of connection to context for test subjects in Italy is considered as an important deficiency, the presence of contextual ties is considered very important in Turkey. So, the results show also big differences with the negative and positive conditions. This might arise from the fact that Italians take contextual ties for granted and do not feel a great need for it, while Turkish subjects consider them important when asked positively. On the other hand psychological requirements are more important for Turkish test subjects both for negative and positive conditions as privacy is considered as a key factor in general well-being. This cultural tendency can be easily read in vernacular living environments as well as those of today. (Minucciani, Saglar Onay, 2019)

INTERIOR DESIGN FOR ALL AGE GROUPS

The background research and the results of the case study conducted with different age groups from different cultures show that the understanding of well-being can not be similar for all age groups. This is result of changing mental and physical conditions all through people's life course. We can assume that even the general well-being of people does not change greatly during their life (supporting the findings of the set point theory), the balance between well-being requirements tend to change. But also these changes, as we have seen in the results of the case study, are also related to the cultural setting and cannot be generalised. However the upward and downward tendencies can be noted in different variables of well-being related to the living environments. For example for the context of Turkey, Istanbul we can see an upward tendency for sensory requirements while we note a downward tendency for functional requirements. On the other hand for Italy, Torino we can observe an upward tendency for social and psychological requirements while we see a downward tendency for all the other requirements. This fact invites us to design interiors as flexible spaces, available to be adapted over time and with time.

If interiors are to be adapted to changing well-being requirements during people's lifespan, the concept of flexibility starts with understanding the target user groups and their roles within the specific context. This becomes a key factor in determining the inclusiveness of the design problem. This is why architects and designers discuss priorities of design while dealing with buildings intended for specific age groups like schools and elderly houses. But the question is not that simple. Many living environments and public interiors are also used by people from all age groups and this necessitates a completely inclusive understanding of interior design.

Regarding inclusive understanding of design, we come across a design approach called inclusive design that emerged in the UK and other countries (Coleman, 1994) from collaborations between industry, designers, researchers and educators. It constitutes a framework and growing body of practice within which design practitioners can understand and respond to the needs of diverse users, with the ultimate inspiration of developing products and services that can meet the needs of the whole population (Clarkson, Coleman, Keates, Lebbon, 2003). On the other hand, inclusive design is mostly used and discussed for design products but not for architecture and interior design. Because for environmental design disciplines designing for the specific user groups is already accepted as a design priority. But this necessitates an in-depth understanding of well-being requirements of these different groups, away from predeterminations. The conception of inclusive interior design is to be handled with all the specific requirements that built up well-being taking into account the changing tendencies of requirements related to different age groups. According to Imrie and Hall (2001), in inclusive design, a very important concept is to understand why design solutions can not be useful for some parts of the society, what is special about them and why they don't fulfil the needs of this specific group. Understanding the missing points leads to a better understanding of how the design can become better. In this sense inclusive design is concerned with meaning and context, participative, human oriented, user re-defined to include users, low cost, democratic and heterogeneous (Imrie, Hall, 2001).

The idea of interior space for everyone is not simply thinking about disabilities or weaknesses of certain group of users. In doing so designers can only minimize the negative effect. On the other hand also the positive effect has a great impact on well-being and it is to be maximised. The survey study discussed above also gives completely different results related to positive and negative conditions. In this sense interior design is much more than eliminating obstacles for everyone, it is about giving the necessary tools to users from all specific user groups so that they can build up their well-being. Therefore the promoters of well-being are to be considered as an important part of interior design.

FUTURE RESEARCH DIRECTIONS

Future research about the relationship between well-being and space needs to focus on different dimensions of well-being defined as contextual, functional, psychological, social, sensory, aesthetic and ergonomic requirements in order to see the differences for age groups. In this way it can be possible to see the variances between age groups for each requirement that has impact on general well-being. But, it should also be noted that the results will most probably show differences for societies with different cultural backgrounds and they can not be generalised.

CONCLUSION

Ageing is a process of adaptation to age-related changes, and successful ageing can be defined as successful adaptation, the measure of success being the maintenance of well-being (Family and Community Development Committee, Parliament of Victoria, 1997). Adaptation could also be regarded as coping with age-related changes (Lazarus & DeLongis, 1983; Ranzijn, R., Luszcz M, 1999). Therefore if the necessary conditions are met getting old can be accompanied by well-being and people can be happier after the long run feeling satisfied with the outcomes of their life and accepting it with all the negatives and positives as they both contribute to the whole.

There are many different theories about the relation between well-being and age. Although common sense dictates that well-being might decrease with age, numerous studies show that it does not decrease but it might even increase. We know that well-being is a sum of numerous factors that need to be in balance. It is obvious that with time our bodies and minds change affecting also our relationship with the spaces we live in. The results of the case study have shown that well-being requirements evolve during people's lifespan but the most significant result concerns the differences between cultures. Examining a wide range of well-being requirements, it is found that spatial preferences were consistently different between the two countries (Italy and Turkey), but within each context they varied little with age.

Designing today's living environments necessitate investigating the well-being requirements discussed throughout this paper. Although all of the well-being requirements are important to fulfil the necessary level of well-being, the ways of achieving these goals differ according to different contexts and understandings of users. Therefore further research needs to focus on the relation of well-being and age in different cultural contexts to study the significant role of culture on well-being requirements.

REFERENCES

Al-Janabi, H. N., Flynn, T., & Coast, J. (2012). Development of a self-report measure of capability well-being for adults: The ICECAP-A. *Quality of Life Research: An International Journal of Quality of Life Aspects of Treatment, Care and Rehabilitation*, *21*(1), 167–176. doi:10.100711136-011-9927-2 PMID:21598064

Alkire, S. (2002). *Valuing freedoms*. Oxford University Press. doi:10.1093/0199245797.001.0001

Allen, J. (2008). *Older people and well-being*. IPPR Institute for Public Policy Research.

Andersson, J. E. (2011). Architecture for the silver generation: Exploring the meaning of appropriate space for aging in a Swedish municipality. *Health & Place*, *17*(2), 572–587. doi:10.1016/j.healthplace.2010.12.015 PMID:21317019

Arnett, J. J. (2007). Suffering, selfish, slackers? Myth and reality on emerging adults. *Journal of Youth and Adolescence*, *36*(1), 23–29. doi:10.100710964-006-9157-z

Australian Government Department of Health, Ageing. (2010). *Kids Matter early childhood–overview: framework, components and implementation details*. Canberra: Commonwealth of Australia.

Balint, M. (1955). Friendly expanses—horrid empty space. *The International Journal of Psycho-Analysis*, *36*(5), 225–241. PMID:14391945

Baltes, P. B., & Baltes, M. M. (1990). Psychological perspectives on successful aging: The model of selective optimization with compensation. In P. B. Baltes & M. M. Baltes (Eds.), *Successful aging: Perspectives from the behavioural sciences* (pp. 1–34). Cambridge University Press. doi:10.1017/CBO9780511665684.003

Barcaccia, B., Esposito, G., Matarese, M., Bertolasoc, M., Elvirad, M., & De Marinis, M. G. (2013). Defining quality of life: A wild-goose chase. *Europe's Journal of Psychology*, *9*(1), 185–203. doi:10.5964/ejop.v9i1.484

Bernard, M. (2004). The you can do it! Early childhood education program: a social-emotional learning curriculum (4–6 year olds). Victoria: Australian Scholarships Group.

Blanchflower, D., & Oswald, A. (2008). Is well-being U-shaped over the life cycle? *Social Science & Medicine*, *66*(8), 1733–1749. doi:10.1016/j.socscimed.2008.01.030 PMID:18316146

Brickman, P., Coates, D., & Janoff Bulmann, R. (1978). Lottery winners and accident victims: Is happiness relative? *Journal of Personality and Social Psychology*, *36*(8), 917–927. doi:10.1037/0022-3514.36.8.917 PMID:690806

Bronfenbrenner, U. (1999). Environments in developmental perspective: Theoretical and operational models. In S. L. Friedman & T. D. Wachs (Eds.), *Measuring Environment Across the Life Span* (pp. 3–28). American Psychological Association. doi:10.1037/10317-001

Burton, E. J., Mitchell, L., & Stride, C. B. (2011). Good places for aging in place: Development of objective built environment measures for investigating links with older people's well-being. *BMC Public Health*, *11*(1), 839. doi:10.1186/1471-2458-11-839 PMID:22044518

Clarkson, J., Coleman, R., Keates, S., & Lebbon, C. (2003). *Inclusive design: Design for the whole population*. Springer Verlag. doi:10.1007/978-1-4471-0001-0

Cohen, S., Tyrrell, D. A., & Smith, A. P. (1991). Psychological stress and susceptibility to the common cold. *The New England Journal of Medicine*, *325*(9), 606–612. doi:10.1056/NEJM199108293250903 PMID:1713648

Coleman, J. S. (1994). *Foundations of Social Theory*. The Belknap Press of Harward University Press.

Cosolino, L., & Sprokay, S. (2006). Neuroscience and Adult Learning. In S. Johnson & K. Taylor (Eds.), *The Neuroscience of Adult Learning: New Directions for Adult and Continuing Education* (pp. 11–20). John Wiley & Sons.

Costa, P. T., & McCrae, R. R. (1980). Influence of extraversion and neuroticism on subjective well-being: Happy and unhappy people. *Journal of Personality and Social Psychology, 38*(4), 668–678. doi:10.1037/0022-3514.38.4.668 PMID:7381680

Cross, S., & Markus, H. (1991). Possible selves across the life span. *Human Development, 34*(4), 230–255. doi:10.1159/000277058

Danermark, B., & Ekstrom, M. (1990). Relocation and health effects on the elderly: A commented research review. *Journal of Sociology and Social Welfare, 17*(1), 25–49.

Davis, J. T., Morris, M., & Drake, M. M. (2016). The moderation effect of mindfulness on the relationship between adult attachment and well-being. *Personality and Individual Differences, 96*, 115–121. doi:10.1016/j.paid.2016.02.080

De Neve, J. E., & Oswald, A. J. (2012). Estimating the influence of life satisfaction and positive affect on later income using sibling fixed effects. *Proceedings of the National Academy of Sciences of the United States of America, 109*(49), 19953–19958. doi:10.1073/pnas.1211437109 PMID:23169627

Deci, E. L., & Ryan, R. M. (2008). Hedonia, eudaimonia, and well-Being: An introduction. *Journal of Happiness Studies, 9*(1), 1–11. doi:10.100710902-006-9018-1

Dehan, P. (2007). *L'Habitat des personnes agées, du logement adapté aux EHPAD, USLD, et unites Alzheimer*. Le Moniteur.

Department for Education and Skills. (2007). *SEAL: guidance booklet*. Retrieved from http://nationalstrategies.standards.dcsf.gov.uk/node/65859

Diener, E., Oishi, S., & Lucas, R. E. (2002). Subjective well-being: The science of happiness and life satisfaction. In C. R. Snyder & S. J. Lopez (Eds.), *Handbook of Positive Psychology*. Oxford University Press.

Diener, E., Suh, E. M., Lucas, R. E., & Smith, H. L. (1999). Subjective well-being: Three decades of progress. *Psychological Bulletin, 125*(2), 276–302. doi:10.1037/0033-2909.125.2.276

Durlak, J., Weissberg, R., Dymnicki, A., Taylor, R., & Schellinger, K. (2011). The impact of enhancing students' social and emotional learning: A meta-analysis of school-based universal interventions. *Child Development, 82*(1), 405–432. doi:10.1111/j.1467-8624.2010.01564.x PMID:21291449

Erikson, E. (1959/1980). *Identity and the life cycle*. Norton.

Fagundes, C. P., Glaser, R., & Kiecolt-Glaser, J. K. (2013). Stressful early life experiences and immune dysregulation across the lifespan. *Brain, Behavior, and Immunity, 27*, 8–12. doi:10.1016/j.bbi.2012.06.014 PMID:22771426

Fagundes, C. P., & Way, B. (2014). Early-life stress and adult inflammation. *Current Directions in Psychological Science, 23*(4), 277–283. doi:10.1177/0963721414535603

Family and Community Development Committee, Parliament of Victoria. (1997). *Report upon the Inquiry into Planning for Positive Ageing*. Victoria, Australia: Parliament of Victoria.

Field, J. (2009). Good for your soul? Adult learning and mental well-being. *International Journal of Lifelong Education, 28*(2), 175–191. doi:10.1080/02601370902757034

Fredrickson, B. L. (1998). What good are positive emotions? *Review of General Psychology, 2*(3), 300–319. doi:10.1037/1089-2680.2.3.300 PMID:21850154

Fredrickson, B. L. (2001). The role of positive emotions in positive psychology: The broaden and build theory of positive emotions. *The American Psychologist, 56*(3), 218–226. doi:10.1037/0003-066X.56.3.218 PMID:11315248

Galambos, N. L., Barker, E. T., & Krahn, H. J. (2006). Depression, self-esteem, and anger in emerging adulthood: Seven-year trajectories. *Developmental Psychology, 42*(2), 350–365. doi:10.1037/0012-1649.42.2.350 PMID:16569173

Gammage, P. (2008). *Learning, engagement and well-being*. Paper presented at the our Children the future, July 2008 Seminar. Early childhood services, Department of Education and Children's Services, Government of South Australia. Retrieved from www.octf.sa.edu.au/files/links/Prof_Philip_Gammage_Learni.pdf

Gutman, L. M., & Feinstein, L. (2008). *Determinants of Aspirations*. Centre for the Wider Benefits of Learning, Institute of Education.

Hanson, J. (2004). The inclusive city: delivering a more accessible urban environment through inclusive design. *Proceedings of RICS Cobra 2004 International Construction Conference: Responding to change*.

Hendry, L. B., & Kloep, M. (2007). Conceptualizing emerging adulthood: Inspecting the emperor's new clothes? *Child Development Perspectives*, *1*(2), 74–79. doi:10.1111/j.1750-8606.2007.00017.x

Hillman, J. (2000). *The force of character and the lasting life*. Random House Publishing.

Imrie, R., & Hall, P. (2001). *Inclusive design: Designing and developing accessible environments*. Spon Press.

Kahneman, D. (1999). Objective happiness. In D. Kahneman, E. Diener, & N. Schwarz (Eds.), *Well-being: the foundations of hedonic psychology* (pp. 3–25). Russel Sage Foundation.

Kansky, J., Allen, J. P., & Diener, E. (2016). Early adolescent affect predicts later life outcomes. *Applied Psychology. Health and Well-Being*, *8*(2), 192–212. doi:10.1111/aphw.12068 PMID:27075545

Kaup, M. L., Kim, H. C., & Dudek, M. (2013). Planning to learn: The role of interior design in education settings. *International Journal of Designs for Learning*, *4*(2), 41–55. doi:10.14434/ijdl.v4i2.3658

Keller, M. L., Leventhal, E. A., & Larson, B. (1989). Aging: The lived experience. *International Journal of Aging & Human Development*, *29*(1), 67–82. doi:10.2190/DEQQ-AAUV-NBU0-3RMY PMID:2628349

Keyes, C. L. (2006). Mental health in adolescence: Is America's youth flourishing? *The American Journal of Orthopsychiatry*, *76*(3), 395–402. doi:10.1037/0002-9432.76.3.395 PMID:16981819

Kuh, D., Ben-Shlomo, Y., Lynch, J., & Hallqvist, J. (2003). Power C. Life course epidemiology. *Journal of Epidemiology and Community Health*, *57*(10), 778–783. doi:10.1136/jech.57.10.778 PMID:14573579

Küster, C. (1998). Zeitverwendung und wohnen im alter [Use of time and housing in old age]. In Deutsches Zentrum für Altersfragen (Ed.), Wohnbedürfnisse, Zeitverwendung und soziale Netzwerke älterer Menschen. Expertisenband 1 zum Zweiten Altenbericht der Bundesregierung [Housing needs, use of time, and social networks of older adults] (pp. 51-175). Frankfurt/Main: Campus.

Lawton, M. P. (1994). Social ecology and the health of older people. *American Journal of Public Health*, *64*(3), 257–260. doi:10.2105/AJPH.64.3.257 PMID:4811769

Lawton, M. P., & Yaffe, S. (1970). Mortality, morbidity and voluntary change of residence by older people. *Journal of the American Geriatrics Society*, *18*(10), 823–831. doi:10.1111/j.1532-5415.1970.tb02834.x PMID:5457319

Layard, R., & Dunn, J. (2009). *A good childhood: Searching for values in a competitive age*. The Children's Society.

Lazarus, R. S., & DeLongis, A. (1983). Psychological stress and coping in aging. *The American Psychologist*, *36*(3), 245–254. doi:10.1037/0003-066X.38.3.245 PMID:6870040

Lipina, S. J., & Colombo, J. A. (2009). *Poverty and brain development during childhood an approach from cognitive psychology and neuroscience*. American Psychological Association. doi:10.1037/11879-000

Lucas, R. (2007). Adaptation and the set point model of subjective well-being. *Current Directions in Psychological Science*, *16*(2), 75–79. doi:10.1111/j.1467-8721.2007.00479.x

Lyubomirsky, S., Sheldon, K. M., & Schkade, D. (2005). Pursuing happiness: The architecture of sustainable change. *Review of General Psychology*, *9*(2), 111–131. doi:10.1037/1089-2680.9.2.111

Mashford-Scott, A., Church, A., & Tayler, C. (2012). Seeking children's perspectives on their well-being in early childhood settings. *International Journal of Early Childhood*, *44*(3), 231–247. doi:10.100713158-012-0069-7

McLennan, W. (1997). *Mental health and well-being: Profile of adults*. Australian Bureau of Statistics.

Miller, G. E., Chen, E., & Parker, K. J. (2011). Psychological stress in childhood and susceptibility to the chronic dis-eases of aging: Moving toward a model of behavioral and biological mechanisms. *Psychological Bulletin*, *137*(6), 959–997. doi:10.1037/a0024768 PMID:21787044

Minucciani, V., & Saglar Onay, N. (2019). Well-being age and space. *Journal of Urban Design*, *25*(2), 275–286. doi:10.1080/13574809.2019.1667758

Moss, M. S., & Lawton, M. P. (1982). Time budgets of older people: A window of four lifestyles. *Journal of Gerontology*, *37*(1), 115–123. doi:10.1093/geronj/37.1.115 PMID:7053392

Mroczek, D. K., & Kolarz, C. M. (1998). The effect of age on positive and negative affect: A developmental perspective on happiness. *Journal of Personality and Social Psychology*, *75*(5), 1333–1349. doi:10.1037/0022-3514.75.5.1333 PMID:9866191

Myers, D. G., & Diener, E. (1995). Who is happy? *Psychological Science*, *6*(1), 10–19. doi:10.1111/j.1467-9280.1995.tb00298.x

Navarro, D., Montserrat, C., Malo, S., Gonzalez, M., Casas, F., & Crous, G. (2017). Subjective well-being: What do adolescents say? *Child & Family Social Work*, *22*(1), 175–184. doi:10.1111/cfs.12215

Oswald, F., & Wahl, H. W. (2003). Place attachment across the life span. In J. R. Miller, R. M. Lerner, L. B. Schiamberg, & P. M. Anderson (Eds.), *Human Ecology: An Encyclopedia of Children, Families, Communities, and Environments*. ABC-Clio.

Peace, S., Holland, C., & Kellaher, L. (2005). Making space for identity. In G. J. Andrews & D. R. Phillips (Eds.), *Ageing and place: perspectives, policy, practice. Routledge Studies in Human Geography, 9*. Routledge.

Pearlin, L. I. (1994). The study of the oldest-old: Some promises and puzzles. *International Journal of Aging & Human Development*, *38*(1), 91–98. doi:10.2190/5VE2-87HW-LR0C-LFDH PMID:8144262

Pearlman, R. A., & Uhlmann, R. F. (1991). Quality of life in elderly, chronically ill outpatients. *Journal of Gerontology*, *46*(2), 31–38. doi:10.1093/geronj/46.2.M31 PMID:1997570

Perroud, N., Paoloni-Giacobino, A., Prada, P., Olié, E., Salzmann, A., Nicastro, R., & Dieben, K. (2011). Increased methylation of glucocorticoid receptor gene (NR3C1) in adults with a history of childhood maltreatment: A link with the severity and type of trauma. *Translational Psychiatry*, *1*(12), e59. doi:10.1038/tp.2011.60 PMID:22832351

Philips, D. R. (2005). *Aging and place. Perspectives, policy, practice.* Routledge.

Pruchno, R. A., Dempsey, N. P., Carder, P., & Koropeckyj-Cox, T. (1993). Multigenerational households of caregiving families: Negotiating shared space. *Environment and Behavior*, *25*(3), 349–366. doi:10.1177/0013916593255003

Ranzijn, R., & Luszcz, M. (1999). Acceptance: A key to well-being in older adults? *Australian Psychologist*, *34*(2), 94–98. doi:10.1080/00050069908257435

Regnier, V. (2002). *Design for assisted living. Guidelines for housing the physically and mentally frail.* Wiley & Sons.

Rowles, G. D. (1978). *Prisoners of space? Exploring the geographical experience of older people*. Westview Press.

Rowles, G. D., Oswald, F., & Hunter, E. G. (2004). Interior living environments in old age. In H. W. Wahl, R. Scheidt, & P. G. Windley (Eds.), *Aging in context: Socio-physical environments (Annual Review of Gerontology and Geriatrics, 2003)* (pp. 167–193). Springer.

Ryff, C. D. (1989). In the eye of the beholder: Views of psychological well-being among middle-aged and older adults. *Psychology and Aging*, *4*(2), 195–210. doi:10.1037/0882-7974.4.2.195 PMID:2789747

Saglar Onay, N., & Minucciani, V. (2018). *Well-being framework as a contributor of sustainability. WIT Transactions on Ecology and the Environment, 217*. WIT Press.

Segerstrom, S. C., & Miller, G. E. (2004). Psychological stress and the human immune system: A meta-analytic study of 30 years of inquiry. *Psychological Bulletin*, *130*(4), 601–630. doi:10.1037/0033-2909.130.4.601 PMID:15250815

Sen, A. (1993). Capability and well-being. In M. Nussbaum & A. Sen (Eds.), *The quality of life* (pp. 30–53). Oxford University Press. doi:10.1093/0198287976.003.0003

Shah, H. & Peck, J. (2005, July). Well-being and the Environment. *Nef.*

Sloane, P. D., Mitchell, C. M., Weisman, G., Zimmerman, S., Foley, K. M., Lynn, M., Calkins, M., Lawton, M. P., Teresi, J., Grant, L., Lindeman, D., & Montgomery, R. (2002). The therapeutic environment screening survey for nursing homes (TESS-NH), an observational instrument for assessing the physical environment of institutional settings for persons with dementia. *The Journals of Gerontology: Series B*, *57*(2), 69–78. doi:10.1093/geronb/57.2.S69 PMID:11867668

Smith, A. E. (2009). *Aging in urban neighbourhoods. Place attachment and social exclusion.* University of Bristol. doi:10.2307/j.ctt9qgmn1

Statham, J., & Chase, E. (2010). *Childhood well-being: A brief summary.* Childhood Well-being Research Center.

Suldo, S., Thalji, A., & Ferron, J. (2011). Longitudinal academic outcomes predicted by early adolescents' subjective well-being, psychopathology, and mental health status yielded from a dual factor model. *The Journal of Positive Psychology*, *6*(1), 17–30. doi:10.1080/17439760.2010.536774

Ulloa, B. F. L., Moller, V., & Sousa-Poza, A. (2013). *How does subjective well-being evolve with age?* IZA Discussion Paper No. 7328.

Van Landeghem, B. (2008). *Human well-being over the life cycle: Longitudinal evidence from a 20-year panel.* LICOS Discussion Paper No. 213. Katholieke Universiteit Leuven.

Walker, A. (2005). A European perspective on quality of life in old age. *European Journal of Ageing*, *2*(1), 2–12. doi:10.100710433-005-0500-0 PMID:28794711

Wallace, J. B. (1992). Reconsidering the life review: The social construction of talk about the past. *The Gerontologist*, *32*(l), 120–125.

Zaff, J., & Sloan Devlin, A. (1988). Sense of community in housing for the elderly. *Journal of Community Psychology*, *26*(4), 381–398. doi:10.1002/(SICI)1520-6629(199807)26:4<381::AID-JCOP6>3.0.CO;2-W

KEY TERMS AND DEFINITIONS

Anxiety Disorders: Feelings of tension, distress or nervousness. Includes agoraphobia, social phobia, panic disorder, generalized anxiety disorder (GAD), obsessive-compulsive disorder (OCD), and post-traumatic stress disorder (PTSD).

Mental Disorder: According to the ICD–10 Classification of Mental and Behavioral Disorders, a disorder implies ‘the existence of a clinically recognizable set of symptoms or behavior associated in most cases with distress and with interference with personal functions’ (WHO, 1992, p. 5). Most diagnoses require criteria relating to severity and duration to be met.

Prevalence: The number of cases of a disease present in a population at a given time.

Chapter 7
Well-Being and Home Environments

ABSTRACT

Home has a special value for human as it reflects the ways of feeling, thinking, and living of their inhabitants. In this sense, home environments have primary importance in individual and social well-being. This chapter aims to discuss the effects of each well-being requirement in home environments using the design framework for well-being in interiors discussed in Chapter 5. In this sense, each requirement is evaluated in terms of its specific value for home, and the discussion is enriched by a survey study that includes questions about contextual, functional, psychological, social, ergonomic, aesthetic, and sensory requirements addressed in the framework.

INTRODUCTION: WELL-BEING IN HOME ENVIRONMENTS

Today the marketing value of a housing unit is mainly determined by its location, its area and its number of rooms and services. On the other hand there are many other factors that determine their value in use. Value in use is mostly related to inhabitants' ways of living and habits related to the living environments. This also has a great effect on the general subjective well-being of inhabitants which is related to contextual, functional, ergonomic, social, psychological, aesthetic and sensory requirements (Saglar Onay, Minucciani, 2018). These requirements can be subject to great changes according to user's different cultural backgrounds. According to a recent research completed

DOI: 10.4018/978-1-7998-4231-6.ch007

by Minucciani and Saglar Onay (2018), well-being requirements are greatly effected by culture. The results of the interview run in Italy and Turkey show that both general tendencies and those related to age change significantly for the two different cultural contexts. Departing from this argument, it can be assumed that standard architectural and interior solutions have negative effects on inhabitant's general well-being. Thus, well-being requirements related to space are not standard. In academic studies and research the architectural rigidness of housing units is mostly discussed through case studies. For example, in the housing literature of Turkey, the uniformity in the quality of architectural design is undertaken as a quality problem, because it causes to monotonous built-environment and inhibits the variety in spatial experience of the inhabitants. Moreover, most of these studies agree generally on that this problem mainly affects users' living standards and decreases their living quality (Tekeli 2008, Sey 1994, Bilgin 2002, Saglar Onay, Garip, Belek Teixeira, 2016)

Not only should a home provide an indoor environment which enables physical well-being, it should also promote mental well-being, providing its residents with a general feeling of happiness and empowerment. There is much overlap between the design features required for these two aspects. Evans et al. (2003) highlight psychosocial processes such that poor housing conditions might impact upon feelings of identity (with consequences for self-esteem), insecurity and risk (with consequences for anxiety and fear), and control (with consequences for self-efficacy). A later review (James 2008) similarly expounded on the impacts of poor housing in terms of powerlessness and stress, lack of privacy, reduced neighbourliness, lower self-image and lower confidence (see Halpern 1995, Hooper et al. 2007, Evans and Stecker 2004 and Cooper-Marcus 1995 for the underlying arguments). Psychosocial benefits from the home would therefore include not only aspects of autonomy and control, but also aspects of security and retreat, and aspects of status and identity (Kearns et al., 2011).

It has been argued that humans have an inherent need for territory and that the possession and ownership of interior spaces satisfies a primal need we share with most mammals (Ardrey, 1966). The interior space of home, in particular, may become a place of territorial centering, the locus from which we venture forth and to which we return (Buttimer, 1980; Rubinstein, 1989). Home may be a place in which the individual may experience a sense of ownership and control (Rutman & Freedman, 1988). The interior space of home may become a place of safety and security. Indeed, a person's residence may become sacred, a locale in which he or she feels protected and shielded from a profane world beyond (Bachelard, 1969; Buttimer, 1980; Eliade, 1959; Rowles, Oswald, Hunter, 2004).

EVALUATION OF WELL-BEING REQUIREMENTS IN HOME ENVIRONMENTS: A CASE STUDY IN ISTANBUL

In the following part of the chapter, home environments will be evaluated using the "well-being framework for interiors" (Figure1) in Chapter 5, "A design framework for well-being in interiors". The well-being framework for interiors can be used both as an evaluation method for existing environments and a roadmap for designing new living environments that can promote well-being. The structure of the framework is based on contextual, functional, psychological, social, sensory, aesthetic and ergonomic requirements, which are always in relation to each other. Connection to context has an important role as the factor that makes every interior unique with its location and relationship to the rest of the world. Functional requirements are concerned with creating spatial solutions that support human activities. Psychological requirements are more indirect as they deal with the outcomes of the interaction between human and space. Social requirements deal with the spatial opportunities for the level of interaction between people. Sensory requirements are concerned with both quantitative factors like heating, illumination, acoustics, sound control etc. and their reflections on human senses. Aesthetic requirements are also important as they refer to the integrity and harmony of all spatial elements that can be pleasing to human senses and can promote well-being. Ergonomic requirements are more concrete and they refer to the relationship between all interior components and human body and capabilities.

Through the chapter first of all each requirement will be discussed in terms of its specific value for well-being in home environments including all aspects that have importance in fulfilling that specific requirement. Then these aspects will be included in survey questions in order to evaluate user opinions and preferences about the components that may have effect on their well-being. In this sense survey results show subjective and culturally bounded opinions and have value as supplementary data in design studies related to home environments.

Figure 1. Well-being Framework for Interiors

The survey study about the evaluation of home environments is realised by using a questionnaire structured to evaluate the relationship between well-being and spatial requirements in the well-being framework. The questionnaire is composed of 3 parts focusing on static information including age, gender, work, family structure and spatial requirements focusing on spatial

Table 1. Structure of the questionnaire

Static information	Age, gender, work, family structure	
Spatial requirements ê Preferences	Connection to context	· Preference of location · Preference of relationship to context
	Functional requirements	Evaluation of importance for spaces with different functions in living environments
	Ergonomic requirements	Evaluation of the importance of ergonomic solutions related to basic activities in living environments
	Psychological requirements	Hierarchy of psychological needs in relation to living environments
	Social requirements	Evaluation of the positive effect of certain social contact types in living environments
	Sensory requirements	· Preference of materials in living environments · Evaluation of the sensory effect of certain environmental and spatial factors in living environments
	Aesthetic requirements	· Evaluation of the positive effect of certain interior elements in living environments · Evaluation of the importance of certain interior elements in living environments

preferences (Table 1). Static information helps to understand the differences between different groups with specific properties. Questions about spatial requirements aim to understand user's preferences and priorities about each requirement. For each question answer possibilities representing nuances for the same requirement are given and users are asked to put these answers in order according to their priorities.

Survey studies were conducted in Istanbul (Turkey) and the questionnaire was applied to totally 60 people, from different age groups, 17 from age group 18-30, 27 from age group 30-60 and 16 from age group 60 and over. The selection of subjects was determined in order to include different social layers from different parts of the city's population including the city centre, commercial areas, education districts and peripheral residential areas. The selection of test subjects from different ages and different layers of the city was evaluated as an important decision that will increase the widespread effect of the research.

The results were evaluated by using the semantic differential ranging from 1 to 3 or 1 to 5 depending on the number of existing answer possibilities for each question. For example for the question about functional requirements which has 12 answer possibilities, only the first 5 preferences were taken and they were given points from 1 to 5 according to their order. The first choice was given 5 points, the second choice 4 points etc. At the end the sum of points for each answer given by all 60 test subjects were represented in the diagrams. Therefore the diagrams signify a tendency of preferences according to the order of answers.

Connection to Context

Connection to context is the most important design criteria that creates architectural and spatial identity by differentiating buildings/spaces with context related data. Contextual ties in interiors can be discussed in two basic steps including the location of that space and its relation to the surrounding natural and built environment. The questions about connection to context are arranged in order to understand the consequences of these two basic aspects. While location is an input that we need to consider in all our design decisions, relationship to the surrounding environment rises as a more controllable issue for designers and architects.

The first question was arranged to understand the preferences of respondents about the location of their home environments. The answer possibilities

Table 2. First question about preference of location

Question: Which of the following environments would you prefer to live? Please put in order	
Natural setting	Urban setting
Near to the sea Near a lake, river Countryside In a mountain village	In the city center near a mall In the city center near a lively pedestrian street In the city center near entertainment/sport areas

were structured in order to include both natural (sea, lake, mountain) and urban attractions. The test subjects were asked to put in order answers each representing a different type of environmental settings for their homes.

For the first question, many of the test subjects expressed their willingness to create strong ties with the water element with the dominance of sea on lake and river. On the other side other natural setting like mountain village and countryside had considerably lower percentage while compared to sea and lake. It also needs to be considered that the water element can represent different values for different cities and cultures. Thus, for the case of Istanbul, being near to the sea represents a great added value as Istanbul is characterised by the Bosporus, one of the world's most beautiful natural straits characterizing the continental boundary between Europe and Asia.

The second question was arranged to understand the effect of certain aspects related to the relationship between living environments and their specific context. Each answer represented different dimensions of contextual ties.

Figure 2. Preference of home location

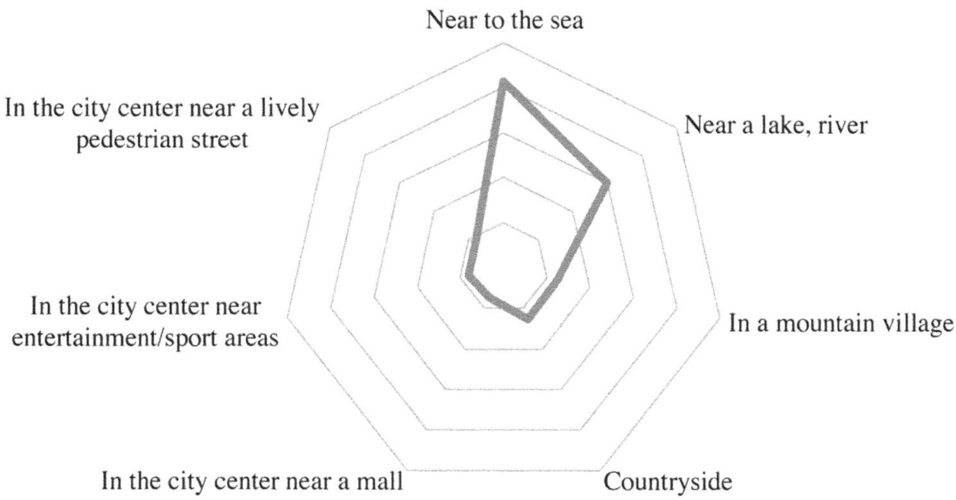

For example, having easy access to interiors and having outdoor extensions were regarded as aspects related to functional requirements while having large

Table 3. Second question about preference of relationship to context

Question: Which of the following make you feel better in an interior environment? Please put in order.	
Having easy access to the interiors.	Functional requirement
Having large openings (windows) with a beautiful view.	Aesthetic, psychological requirement
Having outdoor spaces like terrace, balcony etc.	Functional requirement
Being able to see the sky from roof windows.	Psychological requirement
Having natural light and ventilation	Sensory requirement
Being visible from outside when desired	Psychological, social requirement

openings with beautiful view was considered as an aesthetic and partially psychological requirement (Table 3).

The second question considers different nuances of contextual ties like aesthetic, functional, psychological and sensory dimensions. Having large openings with beautiful view was regarded as the most important positive effect on well-being. The existence of a beautiful view can be considered as the aesthetic dimension of contextual ties. The fact that being able to see the sky is not considered as an important contribution also confirms the importance of view. On the other hand, we can see that a real connection between the interior and the exterior was appreciated. This can be clearly understood

Figure 3. Preference of relationship to context

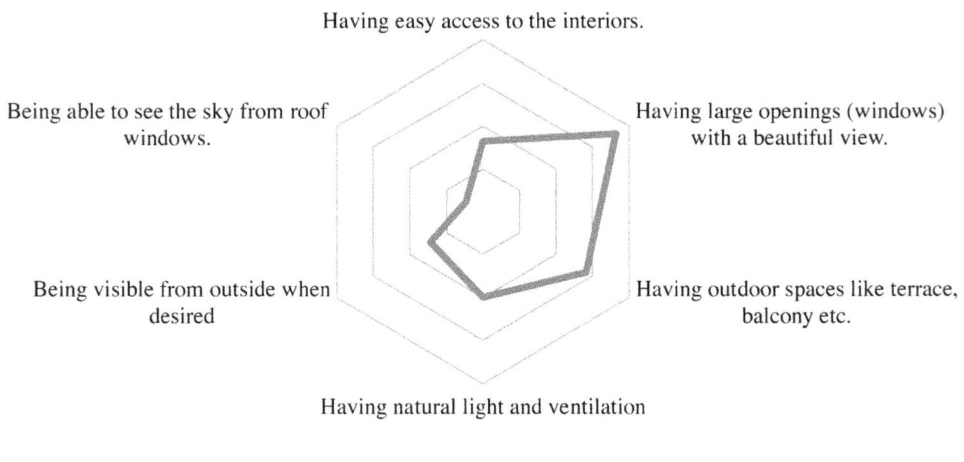

from the desire to have outdoor spaces like terraces and balcony. The sensory dimension of contextual ties is related to the answer “having natural light and ventilation”. This aspect was also considered to be significantly important. Having easy access to interiors was not regarded as important as expected which can be interpreted as a more functional requirement. On the other hand tendency to have large openings and outdoor spaces confirms the necessity of a certain spatial continuity between interior space and the surrounding environment.

Functional Requirements

In general terms functional requirements can be regarded as the basic needs in living environments intended for a certain purpose and they are shaped by the main activities that take place in and around buildings. For a certain interior, the functional requirements may effect the organization of all interior elements such as furniture, lighting etc. The most efficient way of understanding functional requirements is to understand potential user’s intended activities and their way of performing these activities. For multi purpose spaces, functional requirements need to be handled with a flexible approach, which give the possibility to use the same space for different purposes at different times or even for different user profiles. According to Becker (1977), functions related to housing are shaped by activities and meanings within the housing units and every member needs to be satisfied with the housing functions. It is essential to estimate the consistence between spatial mode, family values, needs and activities while determining the design principles of housing units (Kemeny, 1992). Studies in this context are centered on families, households, lifestyles and activities within housing units (Deklava et.al., 2002; Moser, 1987; Sugiyama et.al.2009; Niit, 1993).

Regarding the survey study, functional requirements were associated with spaces with different functions and test subjects were asked to put these spaces in order according to their importance. Although functions may not always correspond to a certain space, the use of spaces rather than functions themselves was regarded as a more comprehensible approach for the questionnaire.

Orderly the living room, kitchen, bedroom and bathroom were regarded as the most important spaces of living environments corresponding to the main domestic activities of sitting, eating, cooking, sleeping, self and body care. On the other hand, spaces like storage room, entrance hall and working

Table 4. The question about evaluation of importance for spaces with different functions in living environments

Question: Which of the fallowing spaces are more important in your living environment? Please put in order.	
Entrance hall	Kitchen
Dining room	Bedroom
Living room	Bathroom
Second/third bathroom	Storage room
Studio for working/studying	Hobby room
Guest room	Garage
Other	

space were not regarded as important as expected due to the fact that the activities related to these spaces (working, storage etc.) can be also fulfilled in other spaces like living room and bedroom according to the needs of users. So the answers to the question about functional requirements tends to show which spaces are indispensible in living environments and represents the need for spatial correspondence only for basic human needs. This can be also interpreted as a tendency towards multifunctional spaces that can change

Figure 4. Evaluation of importance of spaces with different functions in living environments

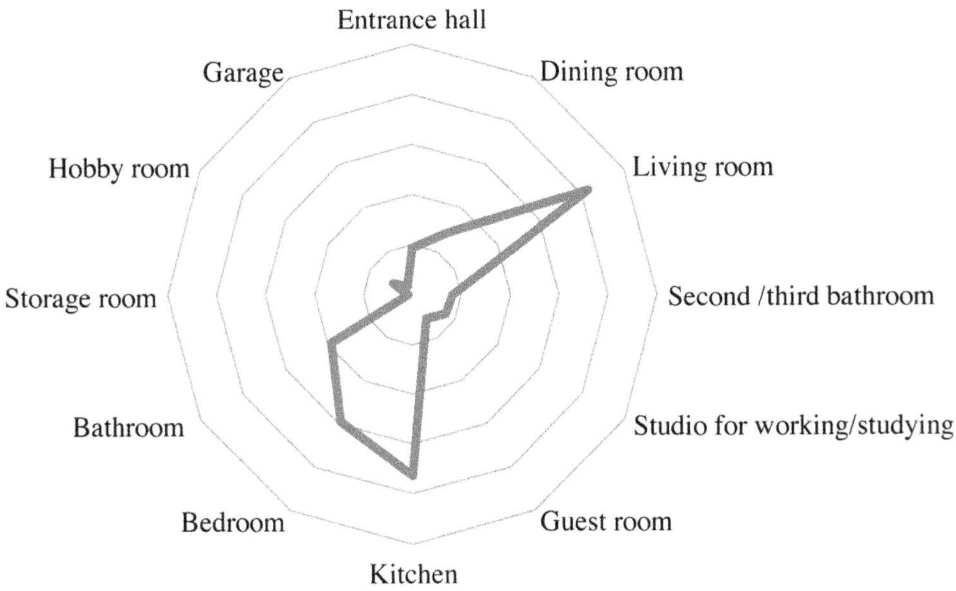

with the changing needs of users related to different activities that take place in different times of the day, week etc. In this sense the living room can be regarded as the most important space in order to fulfill different functional requirements important for the well-being of inhabitants.

Ergonomic Requirements

Ergonomic requirements in interiors are related to the connection between human and space, mainly including physical and cognitive aspects, which help to fulfil a certain task. While physical ergonomics relate to physical activity mainly concerned with human anatomical, anthropometric, psychological and biomechanical characteristics, cognitive ergonomics focuses on mental processes regarding interactions among humans and surroundings (Salvendy, 2012). In this sense, user's capabilities and limitations in seeking to ensure the necessary task define ergonomic requirements.

Wilson (2000) discusses that in the past decades ergonomics would apply anatomical data to examine reach into various appliances for different ages of residents, or else information from experimental psychology to predict how people would respond to different displays on their appliances but life in the home of the future is likely to be enhanced by an array of digital and other telecommunications devices providing our information, entertainment and communication links and by intelligent, adaptive computers allowing supervisory control over domestic sub-systems. This technology will operate in the context of fluid family and social structures, and changing housing, employment and transport needs and policies, which reflect ageing populations, regeneration of city centres, and cottage and networked enterprises. The interactions within this complex sociotechnical system will be the critical focus for the understanding of ergonomics. Therefore, ergonomics is now concerned with the interactions between person, task, equipment and environment (Wilson, 2000).

Regarding the survey study, evaluating the need for ergonomic solutions was associated with the basic activity sets in order to attract attention to the fulfillment of the activity rather than objects and equipment used for the activity. The answers also represented a correspondence between spatial functions and activities and the basic functional spaces were associated with basic activity sets including kitchen, living room, dining room and bedroom.

Having a comfortable kitchen with ease of cooking, storing goods etc. was regarded as the most important requirement. Then followed having a

Table 5. Evaluation of the importance of ergonomic solutions related to basic activities in living environments

Question: Which of the fallowing is more important in your living environment? Please put in order.
Having a comfortable bed
Having a comfortable kitchen with ease of cooking, storing goods etc.
Having a comfortable big living room with plenty of sitting areas
Having a space very comfortable for watching TV, film etc.
Having a space very comfortable for eating with friends and family
Having a separate comfortable, silent room for working/studying
Having a big bathroom with all the necessary accessories
Having plenty of comfortable storage areas in all rooms for necessary household goods

comfortable sitting area, having a space comfortable for eating with friends/ family and having a comfortable space for sleeping. Having comfortable solutions for watching TV, working and the bathroom were considered of secondary importance while having plenty of comfortable storage areas was considered to be the less important requirement. Although ergonomic solutions for storing household goods is one of the most important concerns in home

Figure 5. Evaluation of importance of ergonomic solutions related to basic activities in living environments

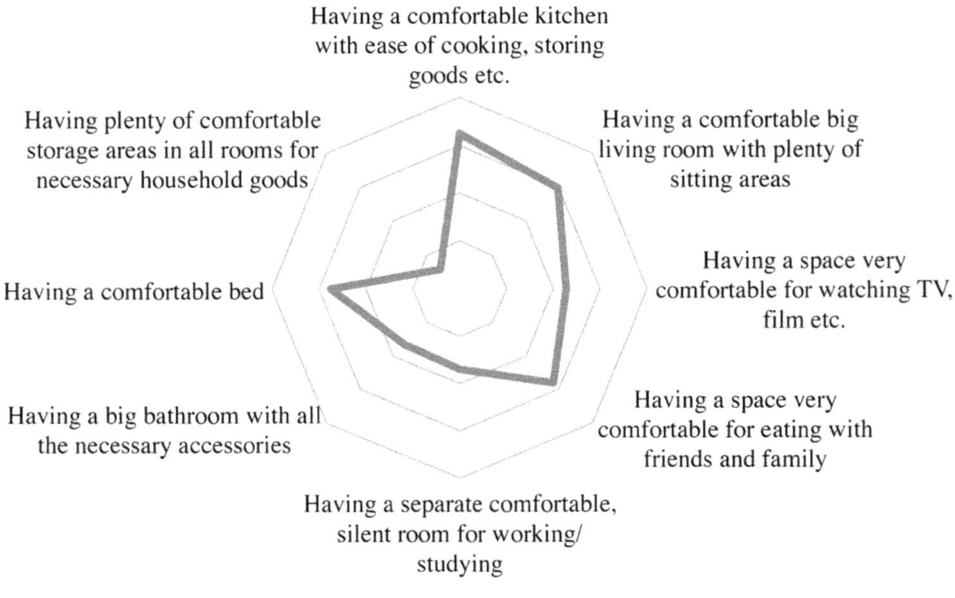

environments, ergonomic requirements related to basic body functions like sitting, eating and sleeping were regarded as much more important.

Psychological Requirements

Theorists ranging from Carl Jung (e.g., 1963) to Clare Cooper-Marcus (e.g., 1995) have proposed that of all places, the home has a particularly powerful symbolic and psychological significance. Homes are worth studying because they are consequential real-world settings in which basic psychological processes are regularly played out. Graham, Gosling and Travis (2015) discuss that homes also provide a consequential real-world context in which to study processes of identity expression and identity development. In this sense they undertook a preliminary study to map out the desired psychological ambiances of a home. The survey consisted of a list of 18 rooms or spaces one might find in a home (e.g., entryway, kitchen, living room, garage, backyard). The instructions were as follows: “As you enter each of the following spaces, what are the most important emotions or perceptions you would like to evoke within yourself and others?” For each of the 18 rooms or spaces, individuals were instructed to select two ambiances from a list of 29 options (e.g., organization, privacy, productivity, rejuvenation, romance, safety, tranquillity, wealth). There was generally strong consensus regarding the ambiances appropriate for each room. As a whole, the results demonstrated that individuals desire a broad range of specific ambiances in their ideal living spaces. The ambiances also differed systematically across spaces. For example, many participants wanted a romantic bedroom, but none of them wanted a romantic garage. These links between ambiances and environments point to the idea that, in addition to having various physical qualities that help or hinder pragmatic tasks, spaces also have psychological qualities that almost certainly affect how the spaces are used, how the occupants feel, and what they think about. So, this means that psychological qualities of living environments need to be regarded as requirements for well-being. Moreover these requirements become even more important in homes where psychological processes are more regular than other spaces.

Psychological needs have different dimensions which all have great importance related to well-being in living environments. Among these, competence, relatedness and autonomy are defined as the three basic innate psychological needs by self-determination theorists (Deci, Ryan, 1985, Ryan, Deci, 2000). Competence related to living environment can be defined as having control over the environment and having the ability to change it if there is need. Relatedness can be interpreted as being connected to other people and autonomy can be evaluated as being free to act as desired for one's own life and act in harmony with one's integrated self. Other than these privacy and security need to be added to the framework because of their importance regarding the relation between human and space. Privacy can be defined as having the desired degree of control over contact with others. Security can be defined as being and feeling free from potential harm.

Regarding the survey study about psychological requirements, the

Table 6. The question about hierarchy of psychological needs in relation to living environments

Question: Which of the following is more important in a living environment? Please put in order.	
Privacy	Desired degrees of control over contact with others
Security	Being and feeling free from potential harm
Competence	Having control of what is going around in the nearby
Relatedness	Interacting, being connected to other people
Autonomy	Being free to act as desired for one's own life and act in harmony with one's integrated self.

Figure 6. Hierarchy of psychological needs in relation to living environments

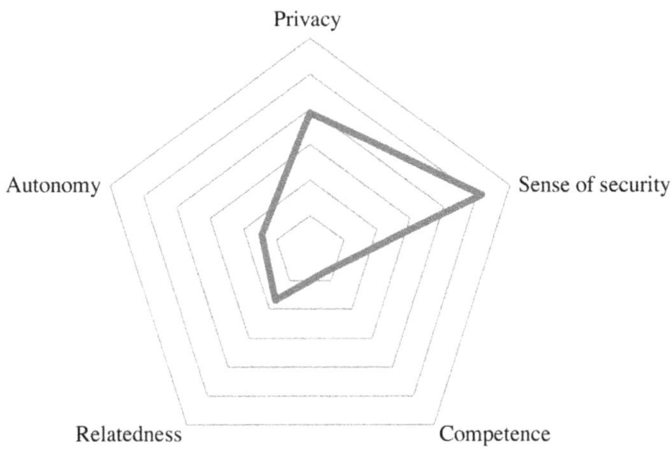

hierarchy of the importance these 5 needs has been evaluated by giving the simple definitions below (Table 6). This was very important in order to avoid misunderstandings about these concepts.

Regarding the results of the survey, the sense of security was the most important psychological need and then followed privacy. Although in general well-being, all these requirements are fundamental, a great hierarchical difference can be noted while focusing on well-being related to living environments. As privacy and security are concepts that have more direct and noticeable ties with space, they were regarded as more important.

Social Requirements

As people live in community, well-being of individuals is related to well-being of others and the interaction between individuals and groups. Keyes (1998) defines social well-being as an individual's self report of the quality of his or her relationship with other people, the neighbourhood and the community.

The effect of social interaction between the family members and close neighbours and relatives can greatly affect the feelings associated to the home environment. According to a qualitative survey applied to a large sample of 250 households in Netherland, the home environment was regarded as most pleasing while communicating with one another and when being accessible to one another. Therefore pleasantness and atmosphere do not occur, they have to be created; that is, they result from action and intentional behaviour. The quality of atmosphere depends essentially on the quality of the interpersonal relations, more specifically on the degree to which people try to approach each other (Pennarz, 1999).

Regarding the effect of social requirements in our living environments, we can think about three different dimensions as the social interaction through collective activities, the communication of social identity and the possibility for social interaction through Internet.

While social interaction through collective activities and communication of social identity require concrete spatial arrangements, social interaction through internet does not have direct relation with space but it represents an alternative way of interaction to the other two dimensions although its effect on well-being can be subject to deep discussion. According to Shklovski, Kiesler and Kraut, (2006) Internet adoption in homes has grown rapidly since the early 1990 and it is a communication technology, with the potential to change peoples' social interaction and one important implication of the Internet's migration to homes and its predominant use for communication is that it could change people's social interaction with their closest ties. According to the first national survey conducted in the United States by Katz & Aspden (1995), internet users had more total contact with family members than non- users, and that they made more new friends, including those they talked with or met on the Internet. On the other hand, Kraut, Lundmark, Patterson, Kiesler, Mukopadhyay & Scherlis (1998) launched a longitudinal study of Internet use in Pittsburgh, Pennsylvania and they reported that participants who used the Internet more showed declines in face-to-face communication with family, smaller social circles, and higher levels of loneliness, stress, and depressive symptoms. The data, the authors suggested, indicated that Internet use can displace valuable time that people spend with family and friends (Nie, Hillygus, & Erbring, 2002). Regarding studies with different and even contrary findings, the effect of Internet on social interaction is not very clear. According to Shklovski, Kiesler and Kraut (2006), the survey studies emerged in the literature make it apparent that the debate over the Internet's social impact had not been decided.

Considering the important but still uncertain role of Internet in social interaction, the question about social requirements have been formulated in order to include answers that are related to social interaction through collective

Table 7. The question about the evaluation of the positive effect of certain social contact types in living environments

Question: Which of the following would make you more happy in your living environment? Please put in order.	
Having spaces suitable to invite friends, neighbours.	social interaction through collective activities
Having a living environment representing your social role/ status/ skills	communication of social identity
Having possibility to connect to world wide web and social networks	social interaction through Internet

activities, the communication of social identity and the possibility for social interaction through Internet (Table7).

The results of the question about social contact types showed that having rooms and spaces suitable to invite friend, neighbours etc is absolutely the most important requirement in living spaces. Communication of social identity was the second important requirement while social interaction through internet was not regarded as important as others. On the other hand, the fact that internet for social interaction is not closely linked to space and spatial requirements

Figure 7. Evaluation of the positive effect of certain social contact types in living environments

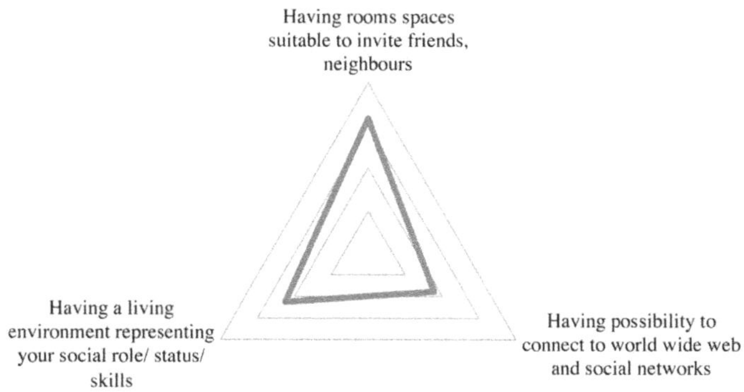

might be the reason for its under evaluation. In this case the insignificance of Internet represents a stronger tendency towards social interaction through collective activities and underlines the importance of spatial solutions that support these collective activities. So the negative tendency of interaction through internet can also be evaluated as a positive tendency towards spatial social interaction.

Sensory Requirements

Sensory requirements in home environments play an important role on well-being. Lighting, ventilation, heating, cooling and acoustics are all issues to be handled with special expertise. Interior designers need to pay attention both on ways to fulfil these requirements and their effects on human senses. A key element of achieving mental well-being is the element of control. When people feel in control of their surroundings and environment, they feel empowerment and stability (Shapiro, 1999), (Taylor et al, 1999). Sensory requirements tend to be controllable through physical arrangements related to thermal environment, acoustics, natural and artificial lighting, humidity and air quality. On the other hand, the effect of these arrangements might be different from person to person depending on personal attributes, activities and sensitivities. Among environmental forces, light and sound contribute significantly, individually and combined, towards the regulation of physio-psychological well-being, being most influential to perception and behaviour.

Natural light is precious for living environments as people enjoy and feel a sense of well-being in daylight and sunlight. Sunlight can also improve the appearance and feel of a home (Green Building Council, 2016). But more than this, daylight has a fundamental role in well-being related to its effect on circadian rhythm. Wirz-Justice and Fournier discuss that daylight is extremely important to synchronise suprachiasmatic nuclei responsible for controlling circadian rhythms that regulates the sleep-wake cycle. Although circadian rhythms are self-sustained, they are adjusted to the local environment by external cues called zeitgebers (time giver), which include light, temperature and redox cycles. One of the basic tenets of chronobiology is that stable synchronisation (known as entrainment) is essential for health and regular 24-hour zeitgeber input is necessary. To permit optimum functioning, therefore, the suprachiasmatic nuclei must first synchronise to the environment and oscillators in peripheral organs must follow. According to Wirz-Justice and Fournier, the most common design example is the customary placing of bedrooms towards the east, so that the orchestration of early morning light conditions within the home can be in phase with the corresponding physiological timing of waking up. It may be beneficial to apply this synchronisation principle to other activities that take place on a regular basis within the home and other buildings, such as food preparation and meals, different kinds of work and educational activities, as well as leisure activities. Moreover, interiors should, as far as possible, be designed in such a way as not to obstruct the entrainment cues offered by the natural environment, particularly in terms of the direction and azimuth of the sun, by minimising overshadowing and allowing distant view. In conditions where insufficient daylight penetration occurs within a building, the design of artificial lighting should seek, as far as possible, to simulate the key characteristics of natural light with respect to those variables that are significant for chronobiology (Wirz-Justice and Fournier, 2010). On the other hand, in case sunlight will feel too intense or hot, and cause discomfort and glare, interiors need to be protected from excess light without blocking its natural effect as synchroniser.

In home environments decisions about materials have significant importance as they can affect all sensory input both physically and psychologically. A change in materials can alter the feeling of a room, colour and the level of illumination as well. While specular materials are highly reflective, matte ones such as natural stone and wood reflect light diffusely equally in all directions. Materials also affect the sense of touch as they have different surface temperatures even when exposed to identical ambient conditions. The surface characteristics of materials also have a great effect on interior acoustics, which is very important in home environments. The qualities and

Table 8. The question about the preference of materials in living environments

Question: Which material do you prefer in living environment? Please put in order			
Stone	Wood	Plastic	Leather
Ceramics	Metal	Paper	Tissue

Table 9. The question about the evaluation of the sensory effect of certain environmental and spatial factors in living environments

Question: Which of the following would annoy you most in your living environment? Please put in order.	
Excess hot or cold	Uncontrolled humidity
Uncontrolled noise	Uncontrolled smell
Uncontrolled airflow	Uncontrolled excess light
Lack of natural light	

textures of materials used in interior space can completely change atmosphere and the way people feel about space.

Regarding the survey study in Istanbul respondents were asked to answer 2 questions. The first question was about the preference of materials in living

Figure 8. Preference of materials in living environments

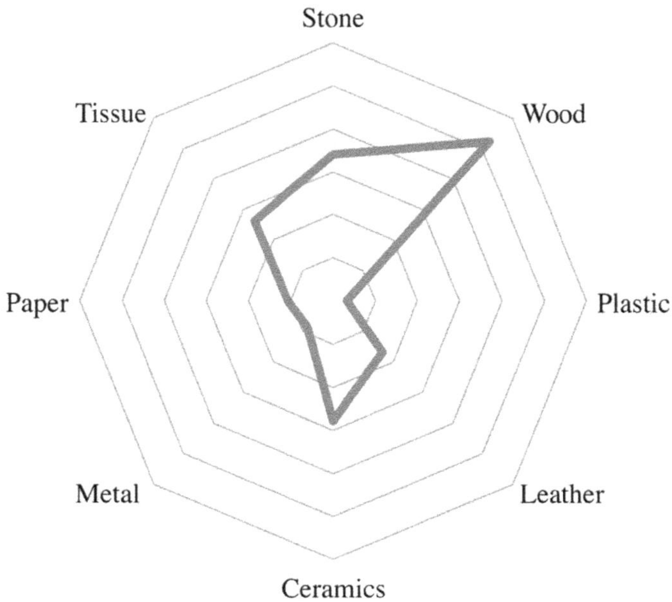

Figure 9. Evaluation of the sensory effect of certain environmental and spatial factors in living environments

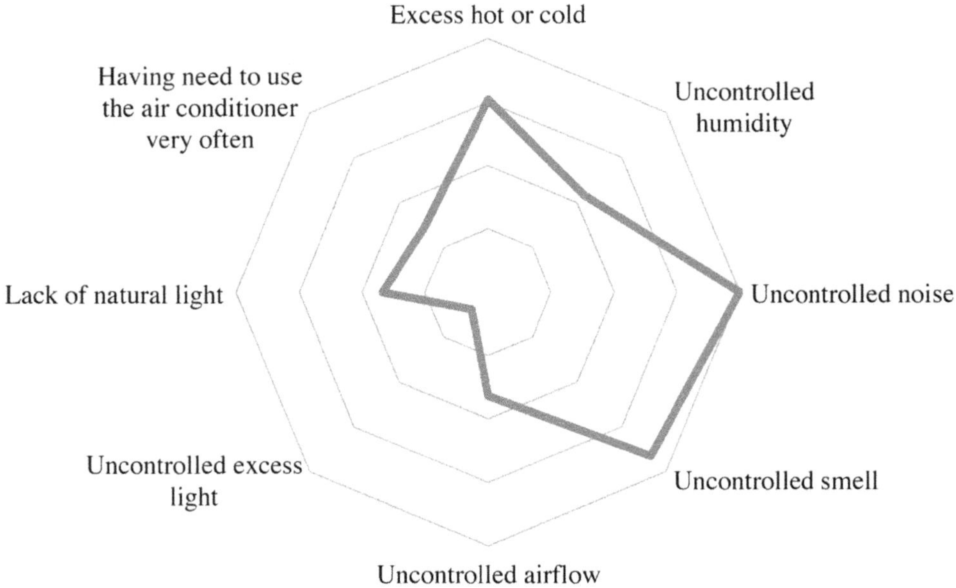

environments. As discussed above, the reason for this is the great effect of materials on the atmosphere of spaces, which can be mostly linked to sensory input. The second question was arranged to understand the sensory effect of certain sensory input, which are also discussed above. Respondents were asked to put in order the effects of sensory input related to negative conditions.

Regarding preference of materials wood has the first place followed by stone, ceramics and tissue.

The results of the evaluation of environmental and spatial factors highlighted the importance of two sensory inputs, which are noise and smell. Test subjects were mostly annoyed by uncontrolled noise and uncontrolled smell and then excess hot or excess cold. Other negative conditions such as lack of natural light, uncontrolled airflow and uncontrolled humidity were not regarded as important as the first three conditions. On the other hand this does not mean that aspects related to natural light, airflow and humidity are not important for well-being as the negative conditions related to only one of these aspects can cause severe health problems. In this sense the evaluations of inhabitants only show their subjective opinions and more immediate responses about the listed aspects, which have negative effect on their understanding of well-being. The real effects can be only evaluated by real world experiences and their long-term outcomes. Because considering wellbeing from the perspective of health sciences, the under evaluated aspects such as uncontrolled humidity and uncontrolled airflow can cause both long term and short term severe health problems.

Aesthetic Requirements

Aesthetic properties related to home environments affect the way people feel depending on spatial aspects, furniture and objects. According to IOM (Institute of Medicine) (2007), in order to be healthy, people need an interesting and aesthetically pleasing environment, sensory stimulation similar to that found in the natural environment. Carlson (2000) discusses that there is a need for aesthetics of the environment, for in such cases our aesthetic appreciation encompasses our surroundings. Going further to interior space, Kendall argues that the way light enters a room, the colours of floors and walls, and even the shape and texture of furniture and home accessories all work together to influence how people feel and how they perform, both consciously and subconsciously. Therefore aesthetics needs to be regarded as a component in our well-being related to space.

The aesthetic quality of space is often associated to its atmosphere, which is defined as characteristic manifestations of the co-presence of subject and object by Böhme (1993). The conception of atmospheres as a phenomenon has its origin in reception aesthetics. Böhme, argues that it is necessary to understand atmospheres not as separate from things and people, they are not 'free floating'. Thus, atmosphere 'is the reality of the perceived as the sphere of its presence and the reality of the perceiver, insofar as in sensing the atmosphere s/he is bodily present in a certain way'. Böhme urges that this principle should be applied to the 'aesthetics of everyday life' as much as to artistic or creative work. In a later essay, he explains further that the 'making' of atmospheres through the idea that making 'does not really consist in producing a thing, but in making possible the appearance of a phenomenon by establishing conditions' (Böhme, 2013). According to Pennarz, atmosphere manifests itself as a double sided process: the atmosphere of a room works on an individual and conversely an individual projects his or her specific mood on the room. According to the qualitative survey applied to a large sample of 250 households in Netherland, the pleasant effect of home environments was associated with three aspects: the arrangement of and connections between rooms, the size of rooms and the form and physical enclosure of

Table 10. The question about the evaluation of the positive effect of certain interior elements in living environments

Question: Which of the following make you feel better in an interior environment? Please put in order.	
Curved walls/ ceilings	Traditional furniture
Contemporary furniture	Different specific colors
Water element	Interior garden with natural plants
Green walls with natural plants	

rooms (Pennarz, 1999). In this sense the manmade potential qualities of architectural characteristics are transformed into "effective environment" on condition of being empowered on the individual conceptual level. Therefore, we can understand the atmospheres as emergent from processes of making. That is from the encounters between people, materials and other elements of the environments of which they are part (e.g. air, light, warmth, scents). Atmospheres are not as such products but they are produced or emergent ongoingly as people improvise their ways through the world (Pink, Mackley, 2016).

As aesthetics and atmosphere are greatly dependent on human experience and moods, it is difficult to define aesthetic requirements. On the other hand, as stated by Böhme it is possible to prepare the necessary conditions that stimulate feelings associated with pleasing atmosphere or at least an atmosphere that can please the potential users.

In home environments the pleasing effect of spatial characteristics is also dependent on the freedom of the users in order to decide on aspects that are dependent on their specific understanding of aesthetics. In a broader design perspective, this also corresponds to visions and notions like continuous design and unfinished design. What makes home territory different from other territories is on one hand the living of the territory and on the other their connection with identity or rather a process of identification of articulation of affect (Wise, 2000).

Regarding the evaluation of aesthetic requirements in living environments, the survey study included two questions. The first question was more about the pleasing effect of certain spatial aspects regarding the overall design of interiors including the decisions about the spatial envelope (walls, ceiling), furniture (traditional, contemporary), colors and the presence of natural elements and aspects (water element, interior garden, green walls) which might contribute to the aesthetic character of interior space.

Figure 10. Evaluation of the positive effect of certain interior elements in living environments

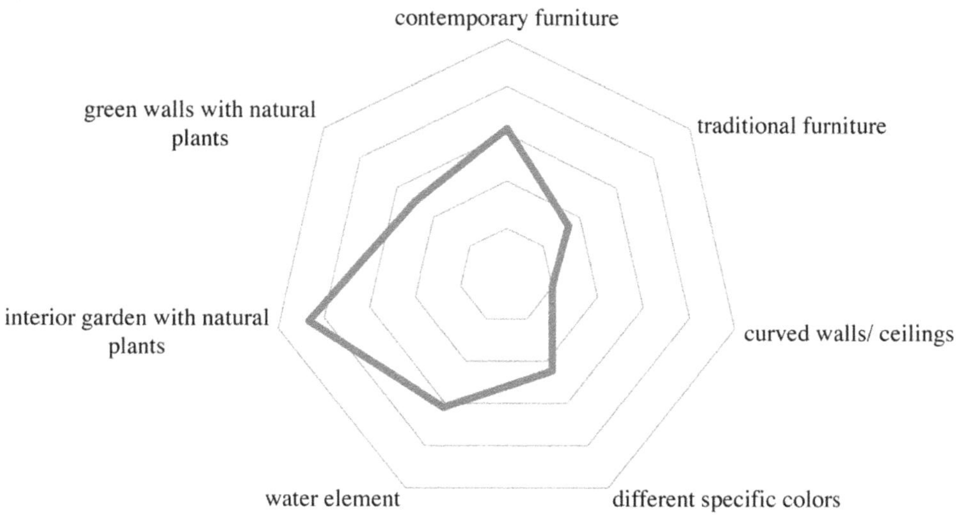

The results about the evaluation of aesthetic requirements regarding the overall design of interiors in living environments were significantly marked by the positive effect of interior garden with natural plants and water element. The presence of contemporary furniture was also regarded as an aspect that can contribute to feeling better in a living environment.

The second question about aesthetic requirements aimed to understand the importance of certain interior components that can contribute to the aesthetic identity of interior space. These components include furniture, objects and accessories that can help to personalise living environments according to one's specific aesthetic sensitivity.

The results about the evaluation of aesthetic requirements regarding the importance of certain interior components was marked by the dominance

Table 11. The question about the evaluation of the importance of certain interior elements in living environments

Question: Which of the following are more important in an interior environment? Please put in order.	
Carpets	Curtains and other textile
Pictures on the walls	Mirrors
The sitting area and sofa	Plants and flowers
Family objects	

Figure 11. Evaluation of the importance of certain interior elements in living environments

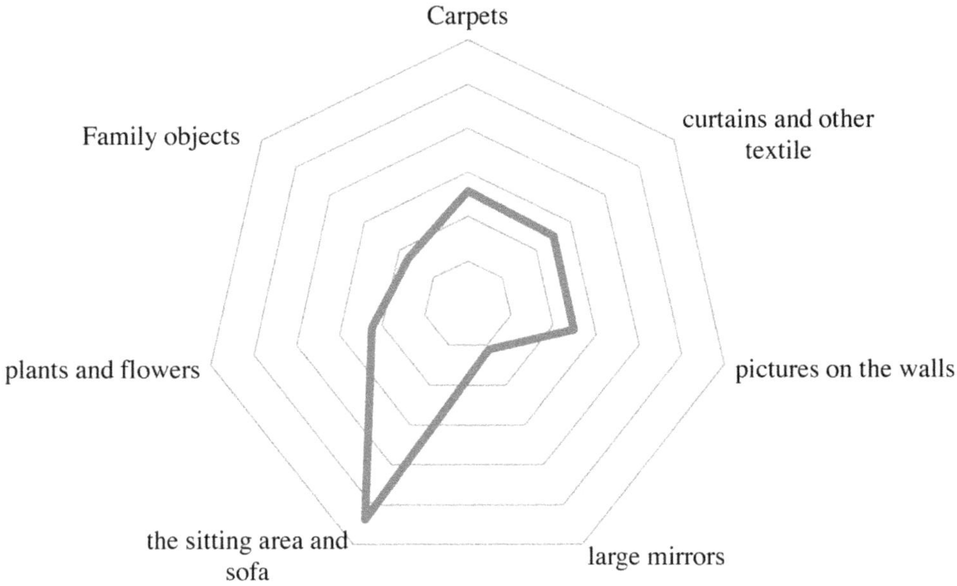

of the sitting area followed by plants and flowers, pictures, carpets, curtains and other textile with similar percentages. On the other hand this significant dominance of sitting area might also be the outcome of its importance regarding functional and ergonomic requirements as the question was not asked only focusing on the aesthetic features of space. Regarding the results of the question about interior components, we can come to an assumption that all components have similar importance (evaluating the dominance of sitting area as a consequence of its relation to functional and ergonomic features).

Results

The survey study about the contextual, functional, psychological, social, sensory and aesthetic dimensions of well-being in living environments can be regarded as an experimental study in order to understand the preferences of users living in a specific context. Table 2 shows the results of the survey study in order to underline the aspects that are considered important for wellbeing in home environments. These results can be used to define the design criteria for home environments at all scales, urban, architectural and interior. At this point it is important to underline that these design suggestions are defined

Table 12. Design suggestions related to wellbeing requirements

Spatial requirements ê Preferences	Connection to context	· Creating strong ties with nature especially the water element · Creating appropriate visual and physical connection between the interior and the exterior · Fulfilling the preference to have large openings (windows) and outdoor spaces like terraces and balconies
	Functional requirements	· Giving special importance to the arrangement of living room and kitchen · Creating multifunctional spaces that can change with the changing needs of users · Arranging dimensions of living rooms in a way that can fulfil different functional requirements
	Ergonomic requirements	· Fulfilling the strong will to have a comfortable kitchen with ease of cooking, storing goods etc. and then a comfortable sitting area (that can also be used to invite friends, neighbours), a space comfortable for eating with friends/family and a comfortable space for sleeping.
	Psychological requirements	· Fulfilling first of all the need for the sense of security and then privacy.
	Social requirements	· Arranging rooms and spaces suitable to invite friends, neighbours etc.
	Sensory requirements	· Giving priority to the use of natural materials such as wood, stone, ceramics and tissue. · Creating solutions to avoid the most annoying sensory input defined as uncontrolled noise, uncontrolled smell, excess hot and excess cold.
	Aesthetic requirements	· Enriching aesthetic quality by connectedness to nature with natural aspects such as interior garden, water element etc. · Giving importance to the arrangement and harmony between the all interior elements.

for the context of Istanbul and they might not be valid for other settings. On the other hand the framework can be applied to different settings in order to define the appropriate design criteria.

It is assumed that the survey study will show different tendencies in another context with different cultural and social backgrounds. For example in a parallel research in order to see user preferences about contextual requirements in different cultural contexts (Minucciani, Saglar Onay, 2019), the same questions were applied to totally 60 people, 20 from age group 18-30, 20 from age group 30-60 and 20 from age group 60 and over. As expected the preferences of users from Istanbul and Turin were not similar but on the other hand general tendencies were not too different. For example for both questions the first preferences were the same. Both test subjects in Italy and Turkey, preferred a living environment near to the sea and they appreciated most having large openings with beautiful view (Figure 12). On the other

Figure 12. Evaluation of the importance of certain interior elements in living environments

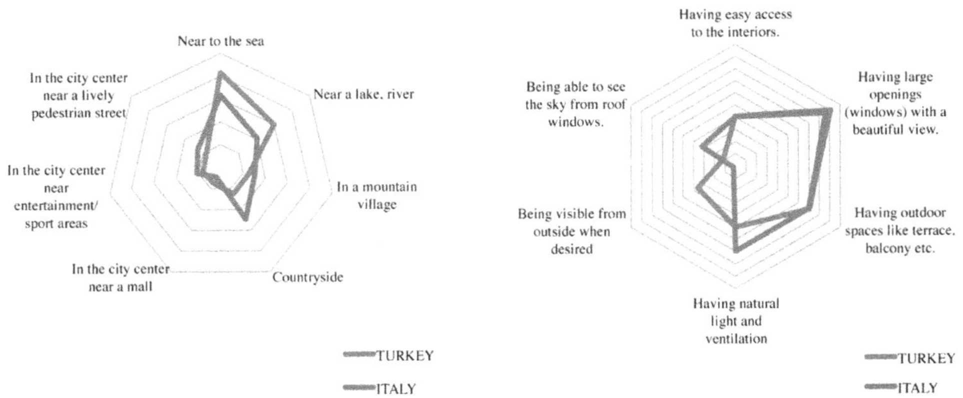

hand, the second preferences for living environments were different (for Italy countryside and for Turkey lake/river). For connection to context the first three preferences were the same while the following three preferences were notably different. This proves that cultural differences have a particular effect on user preferences related to well-being.

As discussed above, the findings show subjective evaluations of inhabitants in a specific context. The difficulty of the research in the field of well-being is the elevated level of subjectivity. While focusing on the effect of the built environment on well-being there is another aspect, which makes the situation even more difficult, is that, the existence of factors that the inhabitants themselves cannot always be aware of or cannot be able to define/ evaluate properly. For example lack of daylight has been selected as the fourth important sensory effect after uncontrolled noise, uncontrolled smell, excess cold and excess hot. On the other hand we know that lack of daylight can cause severe health problems and can be considered as one of the key elements in well-being related to the living environment. So, the environmental designer should be aware of the importance and consequences of each requirement and consider the user evaluations as important supplementary data but not definitive data.

FUTURE RESEARCH DIRECTIONS

Designing today's living environments necessitate investigating the well-being requirements discussed throughout this paper. Although all of the well-being requirements are important to fulfil the necessary level of well-being, the ways of achieving these goals differ according to different contexts and understandings of users. Future research needs to focus on observations and further survey studies that evaluate the effects of spatial requirements in different cultural contexts as cultural differences can completely change the hierarchy of well-being requirements (Saglar Onay, Minucciani, 2019).

CONCLUSION

Well-being in home environments arises both as an environmental and social issue which can have a great impact on the present and the future. The discussion of well-being in home environments through the framework including contextual, social, psychological, functional, ergonomic, aesthetic and sensory requirements aims to create consciousness about all aspects that have effect on well-being. In this sense each requirement have a specific contribution to well-being and must be considered as design input regarding design activities.

Regarding the results of the framework applied to home environments, connection to context can be seen as the starting point and the link to other design scales including architectural and urban scales. One of the most important findings of the survey study is about the strong will to have a certain level of spatial continuity between interior space and the surrounding environment. The tendency to have large openings, terraces, outdoor spaces and a concrete connection with pleasing natural settings such as the sea, lake etc. confirms the importance of spatial continuity in home environments.

The findings about functional requirements show that the spaces related to basic activity sets such as eating, cooking, sitting, sleeping and personal care are regarded as the most important and indispensible components of the home environment. Other activities can be regarded as activities that might be associated with the basic activities. This necessitates the multi-functional use of living environments. In this sense interior spaces answering to basic activities can be designed in order to fulfil other secondary activities, which might be also temporary such as working, playing, exercising etc. The findings of the study can also be interpreted that the living room and the kitchen are interiors that have stronger potential for multifunctional interior solutions.

The results of the survey about ergonomic requirements also confirm the importance of basic activity sets. People see it more important to have a comfortable kitchen with ease of cooking, a comfortable sitting area, a relaxing space for sleeping and having a space comfortable for eating with friends/family.

Regarding psychological requirements of the framework, the survey study underlines the importance of the sense of security and then privacy. These two aspects can be seen as needs that have direct relation with space and this might be reason why they are regarded more important than other innate psychological needs (discussed above as competence, autonomy and relatedness). On the other hand it is essential to underline that the home territory is different and more important than other territories because of its great potential to fulfil the three basic psychological needs. Acting upon ones's own environment gives a sense of achievement and control making home a place for self-expression and for freedom of action. Home also represents refugee from the outside world, a place where one can control the level of social interaction and regulate the level of privacy and interdependence (Meesters, 2009).

The results of the survey about social contact types showed that having spaces suitable to invite friends, neighbours etc. is absolutely the most important social requirement in living environments. Communication of social identity was also regarded as an important need in home environments. Home environment is strongly related to identity and represents the values associated with family and individuals. People refer to their home as a symbol of how they want to see themselves and how they want to be seen by others.

About sensory requirements the results of the evaluation of environmental and spatial factors highlighted the importance of two sensory inputs, which are noise and smell. Then followed excess hot and excess cold. Test subjects were mostly annoyed by factors that have immediate effect on them. On the other hand other factors such as humidity, lack of light, excess light, uncontrolled airflow etc. are also factors that can have short-term and long-term negative effects on health and wellbeing. In this sense designers and architects need to take all these aspects as indispensible components of sensory wellbeing.

Regarding the results of the survey study about aesthetic requirements in general it can be assumed that components recalling connectedness to nature are regarded as the decisive aspects that increase the aesthetic identity of space while the importance of other interior decisions and interior elements like furniture, textile, objects and colours does not show significant differences.

An important aspect about well-being requirements is their elevated subjectivity just like the general concept of well-being. Moreover survey studies on well-being requirements rely on user's immediate evaluations but they don't evaluate the long-term outcomes of conditions related to each requirement. In this sense inhabitants might under evaluate some requirements (like the sensory requirement of daylight) that are extremely important for general health and well-being. So, it is necessary to consider the framework as a road map that will guide research about the environment, user and other aspects that become input for the design process. Consequently, the findings of survey studies in the field of well-being need to be supported by long-term interdisciplinary evaluations which include insights especially from psychology, health studies and environmental design studies including urban, architectural and interior scales.

According to their contribution to well-being, home environments need to be seen as the most powerful promoters of well-being related to space. Architects and designers need to be aware of this potential of homes and work with other disciplines involved in the general research of well-being.

REFERENCES

Ardrey, R. (1966). *The territorial imperative*. Atheneum.

Bachelard, G. (1969). *The poetics of space*. Beacon Press.

Bilgin, İ. (1994). Sıradan olanın yeniden üretimi ve konut sorunu. In *Proceedings of the Turkey Annual Housing Panel*. Ankara: Chamber of Architects.

Böhme, G. (1993). Atmosphere as the fundamental concept of a new aesthetics. *Thesis Eleven, 36*, 113–126.

Böhme, G. (2013). The art of the stage set as a paradigm for an aesthetics of atmospheres. Ambiances. *International Journal of Sensory Environment, Architecture and Urban Space.*

Buttimer, A. (1980). Home, reach and the sense of place. In A. Buttimer & D. Seamon (Eds.), *The Human Experience of Space and Place* (pp. 166–187). St. Martin's Press.

Cooper-Marcus, C. (1995). *House as a mirror of self.* Conari Press.

Deci, E. L., & Ryan, R. M. (1985). *Intrinsic motivation and self-determination in human behavior*. Plenum. doi:10.1007/978-1-4899-2271-7

Dekleva, A., Gatto, M., Gregoric, T., Sedlak, R., & Stroumpakos, V. (2006). *Negotiate my boundary*. Birkhauser.

Eliade, M. (1959). *The sacred and the profane*. Harcourt, Brace & World.

Evans, G. W., & Stecker, R. (2004). Motivational consequences of environmental stress. *Journal of Environmental Psychology*, *24*(2), 143–165. doi:10.1016/S0272-4944(03)00076-8

Evans, G. W., Wells, N. M., & Moch, A. (2003). Housing and mental health: A review of the evidence and a methodological and conceptual critique. *The Journal of Social Issues*, *59*(3), 475–500. doi:10.1111/1540-4560.00074

Fournier, C., & Wirz-Justice, A. (2010). Light, health and well-being: Implications from chronobiology for architectural design. *World Health Design: Architecture, Culture, Technology*, *3*(1), 44–49.

Graham, L. T., Gosling, S. D., & Travis, C. K. (2015). The psychology of home environments: A call for research on residential space. *Perspectives on Psychological Science*, *10*(3), 346–356. doi:10.1177/1745691615576761 PMID:25987512

Green Building Council. (2016). *Health and well-being in homes*. Green Building Council.

Grimwood, C. J., Skinner, C. J., & Raw, G. J. (2002). The UK national noise attitude survey 1999/2000. In *Proceedings of Noise Forum Conference*. Watford: Building Research Establishment.

Halpern, D. (1995). *Mental health and the built environment*. Taylor and Francis.

Hooper, C., Gorin, S., Cabral, C., & Dyson, C. (2007). *Living with hardship 24/7*. Frank Buttle Trust.

Institute of Medicine (IOM). (2007). *Green healthcare institutions; Health, environmnet and economics*. The National Academies Press.

James, C. (2008). *Homes fit for families: An evidence review*. Families and Parenting Institute.

Katz, J., & Aspden, P. (1997). A nation of strangers? *Communications of the ACM*, *40*(12), 81–86. doi:10.1145/265563.265575

Kearns, A., Whitney, E., Mason, P., Petticrew, M., & Hoy, C. (2011). Material and meaningful homes: Mental health impacts and psychosocial benefits of rehousing to new dwellings. *International Journal of Public Health*, *56*(6), 597–607. doi:10.100700038-011-0275-3 PMID:21842221

Kemeny, J. (1992). *Housing and Social Theory*. Routledge.

Kendall, D. (2010, July 21). *Aesthetics and happiness: How space affects well-being*. Retrieved from https://www.huffpost.com/entry/how-to-be-happy

Keyes, C. L. M. (1998). Social well-being. *Social Psychology Quarterly*, *61*(2), 121–140. doi:10.2307/2787065

Kraut, R., Patterson, M., Lundmark, V., Kiesler, S., Mukhopadhyay, T., & Scherlis, W. (1998). Internet paradox: A social technology that reduces social involvement and psychological well-being? *The American Psychologist*, *53*(9), 1017–1031. doi:10.1037/0003-066X.53.9.1017 PMID:9841579

Meesters, J. (2009). *The meaning of activities in the dwelling and residential environment*. IOS Press.

Moser, C. O. N. (1987). Women, human settlements and housing: A conceptual framework for analysis and policy-making. In C. O. N. Moser & L. Peake (Eds.), *Women, human settlements and housing* (pp. 12–32). Tavistock Publications.

Nie, N., Hillygus, S., & Erbring, L. (2002). Internet use, interpersonal relations and sociability: Findings from a detailed time diary study. In B. Wellman (Ed.), *The Internet in everyday life* (pp. 215–243). Blackwell Publishing. doi:10.1002/9780470774298.ch7

Niit, T. (1993). Housing characteristics, family relations and lifestyle: An empirical study of Estonian families. In E. G. Arias (Ed.), *The Meaning and Use of Housing* (pp. 327–341). Avebury.

Pallasma, J. (2005). *The eyes of the skin*. John Wiley &Sons.

Pam, M. S. (2013). Security. In *Psychology Dictionary*. Retrieved from https://psychologydictionary.org/security/

Pennarz, P. J. J. (1999). Home: The experience of atmosphere. In I. Cieraad (Ed.), *At Home: An Anthropology of Domestic Space* (pp. 95–105). Syracuse University Press.

Pink, S., & Mackley, K. L. (2016). Moving, making and atmosphere: Routines of home as sites for mundane improvisation. *Mobilities*, *11*(2), 171–187. doi:10.1080/17450101.2014.957066

Rowles, G. D., Oswald, F., & Hunter, E. G. (2004). Interior living environments in old age. In H. W. Wahl, R. Scheidt, & P. G. Windley (Eds.), *Aging in context: Socio-physical environments (Annual Review of Gerontology and Geriatrics, 2003)* (pp. 167–193). Springer.

Rutman, D. L., & Freedman, J. L. (1988). Anticipating relocation: Coping strategies and the meaning of home for older people. *Canadian Journal on Aging*, *7*(1), 17–31. doi:10.1017/S071498080000708X

Ryan, R. M., & Deci, E. L. (2000). Self-determination theory and the facilitation of intrinsic motivation, social development, and well-being. *The American Psychologist*, *55*(1), 68–78. doi:10.1037/0003-066X.55.1.68 PMID:11392867

Saglar Onay, N. (2017). Analysing spatial potentials of historic buildings. Santarcangelo di Romagna, Rimini: Maggioli Editore.

Saglar Onay, N., Garip, B., & Belek Teixeira, M. (2016). Searching for flexible interior design solutions: Mass housing units in Istanbul. In L. Crespi (Ed.), *Design Innovations for Contemporary Interiors and Civic Art* (pp. 270–284). IGI Global.

Saglar Onay, N., Garip, B. & Garip, E. (2017). A flexible user centered design model for social housing units. *Tafter Journal, 97.*

Saglar Onay, N., & Minucciani, V. (2018). Well-being framework as a contributor to sustainability. *WIT Transactions on Ecology and the Environment*, *217*, 699–710. doi:10.2495/SDP180591

Salvendy, G. (2012). *Handbook of human factors and ergonomics*. John Wiley & Sons. doi:10.1002/9781118131350

Sey, Y. (1994). Konutta kalite ve maliyet. In T. Aktüre (Ed.), Konutta Kalite. Ankara: Mesa.

Shapiro, D. H., & Astin, J. (1998). *Control therapy*. Wiley.

Shklovski, I., Kiesler, S., & Kraut, R. (2006). *The Internet and social interaction*. Carnegie Mellon University.

Sugiyama, T., Thompson, C. W., & Alves, S. (2009). Associations between neighborhood open space attributes and quality of life for older people in Britain. *Environment and Behavior*, *41*(1), 3–21. doi:10.1177/0013916507311688

Taylor, S. E., & Brown, J. E. (1999). Illusions and well being: A social psychological perspective on mental health. *Psychological Bulletin*, *103*, 293–210.

Tekeli, İ. (2008). 1980'li Yıllara Kadar İzmir'deki Konut Sunum Biçimlerinde Yaşanan Çeşitlenmeler. In *Proceedings of İzmir'de 80'li Yıllardan Günümüze Konut ve Mimarlık Kültürü*. Association of Professional Architects of Izmir.

Wilson, J. R. (2000). Fundamentals of ergonomics in theory and practice. *Applied Ergonomics*, *31*(6), 557–567. doi:10.1016/S0003-6870(00)00034-X PMID:11132041

Wise, J. M. (2000). Home: Territory and identity. *Cultural Studies*, *14*(2), 295–310. doi:10.1080/095023800334896

Zumthor, P. (2006). *Atmospheres*. Birkhauser.

KEY TERMS AND DEFINITIONS

Autonomy: Being free to act as desired for one's own life and act in harmony with one's integrated self.

Competence: Having control of what is going around in the nearby.

Flexibility: The ability to adapt to changes.

Privacy: Desired degrees of control over contact with others.

Relatedness: Interacting, being connected to other people.

Security: Being and feeling free from potential harm.

Chapter 8
Well-Being and School Life

ABSTRACT

The "school climate" includes both social and physical aspects of school life and can affect positively (or not) behavior, achievements, and cultural development. The school environment has attracted great attention from architecture, but perhaps the overall well-being of the students has taken second place to educational needs. As in all work environments, to which the school can be partially compared, we need to overcome the concept of well-being as an instrumental element to the best performance of people, in favor of a holistic conception that tends to a better quality of life (which also influences some profitability). This chapter reports and comments on the results of a survey carried out on two different school complexes (with students aged between 11 and 14 years) in the Turin metropolitan area, where different socio-economic and cultural conditions are represented. The questionnaire submitted to the students is articulated according to different requirements of well-being: functional, social, psychological, ergonomic, aesthetic, sensorial, and connection with the context.

INTRODUCTION

Feeling of happiness, enjoyment, curiosity and fulfillment – as well as having good opportunities and experiencing positive relationships - are crucial in the concept of well-being. (Huppert 2008). These are, indeed, important issues for education, throughout life and especially for young people, supporting their personal and social development (Jackson and Norman, 2013).

DOI: 10.4018/978-1-7998-4231-6.ch008

Since school is the place where children and young people spend most of their time, the school climate, is crucial for their present and future well-being. It affects development of the person, not only from the cultural point of view but also from the psychological and social ones. Furthermore, the environment (physical and emotional) has a great influence on school performance and ultimately on the skills and motivation of future adults.

Seligman (2002) states that "there is a difference between the talents with which you are born and that you are therefore automatically good, and the forces that you choose to develop": for this reason, it is important to cultivate and implement our own strengths. This vision is theoretical, according to many critics, as it underestimates the cultural aspects and circumstances over which the individual cannot have control. In any case, school is certainly crucial in this personal and social development, and in particular the influence that the architectural environment can have is not secondary.

For many adults, work is the most important social environment, just as for young people school is. Then it is important to understand the nature of work, which according to White is different in autonomous work (which we chose to do in order to achieve self-realisation) and in the 'heteronomous' one (in which we're engaged because we have to do it). Most jobs and educational contexts require "heteronomous" work (White 2011).

A study by Amabile and Kramer has shown that most people believe that work is also important for self-fulfillment. The study also showed that everyone has an "Inner Work Life", consisting of the continuous emotions and motivations, of different nature, that the work generates in them. People's reactions to this flow of emotional events affect not only their performances but also their well-being. When people have a positive "Inner Work Life", they are more creative and productive, while when they have a poor "Inner Work Life", the opposite happens. The most important factor influencing the "Inner Work Life" is precisely the perception of an increasing significance of the work, which in turn Amabile and Kramer called "progress principle." Their study identified four categories of elements that reinforce the positive aspects of work namely respect, encouragement, emotional support, sense of belonging - and certainly these are factors that can influence the sense of well-being in other social situations as well (Amabile and Kramer 2011).

For youngest people, these are decisive factors: the attitude towards life and the environment evolve and are gradually defined in school age.

Well-being depends on the possibility of achieving goals and exploiting the opportunities offered to us. Nussbaum defined ten capacities, i.e. real opportunities based on personal and social circumstances (Nussbaum 2000): among these he identified the "Senses, Imagination and Thought" (i.e. being able to use the senses, and having sufficient education to best exercise imagination and reasoning), the "Emotions" (in particular, safeguarding one's own emotional development from fear and anxiety), "Other Species" (living in relation to animals, plants, and nature), "Play" (enjoying recreational activities), and "Control over one's Environment" (intended as political environment and material one). In particular, this last point is not particularly explained as architectural context, and this lack should be filled.

Actually there are many pedagogical essays while few studies address the issue of schools starting directly from the well-being needs of the students, although well-being is something specific to different groups: children, employees or other (Jackson 2013).

WELL-BEING, SCHOOL AND ARCHITECTURE

School environment is the first opportunity for socializing, in addition to learning. Children's perception and behavior is affected by the physical environment (shapes and proportions, materials, colors and textures), and in particular children like spaces where they feel safe, and where they can express themselves. Child cognitive development depends on many different genetic and environmental factors (Türel and Gür 2019).

As Merleau Ponty(1945) pointed out, the only way we can know the space around us is our perception. Many scholars have shown that the perceptive behavior of children is different from that of adults, and in any case evolves with time and age. Therefore, their surrounding architectural environment should take this into account.

Several scholars discussed this issue, arguing that child development has to be considered as a whole (Yavuzer, 2000, Gur, 2014), being affected by physiology, psychology, social life, interaction with environment (Piaget and Inhelder 1967, Vygotsky1986, Kozulin, 2003), genetic factors, family structure, culture and origin (Olds, 1979).

In particular, Piaget considers intelligence as a form of biological adaptation, which from time to time builds new mental structures that allow to understand and explain the environment, adapting to it (Piaget 1963 and 1974). He proposes a model in which cognitive structures are built through the continuous interaction between the subject's innate mental processes and the physical and social environment. Compared to Piaget, likewise Vygotskij considers the child as an active builder in the context, but while Piaget refers

Figure 1. Well-being Framework for Interiors
(Saglar Onay, Minucciani 2018).

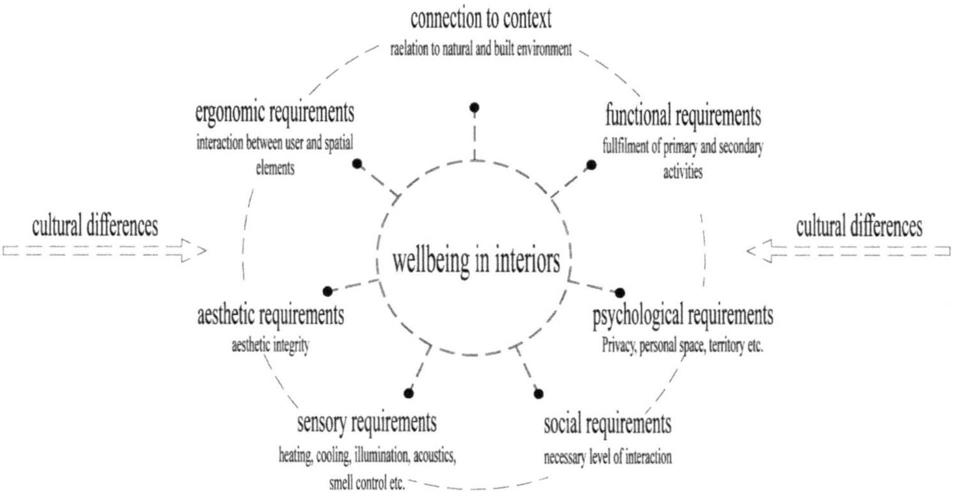

to the physical environment, Vygotskij emphasizes the social one, intended both as culture and as interactions. Then the child must be considered within the dominant culture to which he/she belongs.

During the XX century many architects dealt with this topic in a very broad way, and with specific attention to the pedagogical aspects and well-being of students. In particular, Modern Architecture faced a significant change in lifestyle, which of course was mainly reflected in domestic and work spaces, but the other building types were also affected by these changes. The school environment evolved: health-related requirements - such as lighting conditions or air circulation - made the school not only a healthier place to live, but also "more open and interconnected, with more transparency and spatial complexity – both inside and out. The environment, in general, became more child-friendly" (Hille 2011). At the same time, as Hille underlines, the development of modern educational programs directly involved the quality of architectural spaces. It was recognized that the architectural identity of the school, its connection to the context, the flexibility and variability of spaces, as well as the support to social dimension, were crucial elements for the quality of a school.

Designers of the last century devoted special attention to the school environment. Some of them will be cited here in order to exemplify the decades of the so-called Modern Architecture.

Dudok Architect, as part of a vast and innovative program for the garden city of Hilversum, designed new schools in the 1920s with the aim of "deinstitutionalize and humanize" them, to make them more child-friendly, more architecturally varied and more in accordance with the outdoor contest. The Rembrandt School is an example of these projects, providing spaces of different sizes for large groups as well as small ones, and with interior features unusual for those times, such as brightly colored plastered walls (Hille 2011).

Modern architects, from Eliel Saarinen to Pekins and Will, from Jean and Henri Prouvé to Richard Neutra and Marcel Breuer, paid particular attention to natural light, creating large, bright rooms. They took care of the relationship with outdoor spaces as much as possible, by creating systems of courtyards, patios, porches.

Among the many Modern Architects who dealt with school design it is worth mentioning Hans Scharoun (Burke 2005, Hille 2011), who emphasized the educational development of social relationships between individuals and society and took great care of the "psychological spaces" for children suitable to play but also to promote their spiritual development. The school in Marl and above all the Brothers School in Lünen are particularly notable. In the second one, Scharoun concretized his ideas of a reformed pedagogy centered on the psychology of children and aimed, after the Nazi experience, at an education to democracy. The Lünen school is conceived as a city with irregularly shaped environments for the social life of the students, with different areas for public and private use. The heart of the architecture is the so-called Klassenwohnung, the 'class-house', consisting of a classroom, a space for group work, an entrance with coatroom and an outdoor space; each 'class-house' is also equipped with a sophisticated system of ventilation and air heating that the teacher could adjust individually through a control panel. During the renovation, a big surprise was the rediscovery of the lively polychromy with strong tones on the outside and delicate pastel colors on the inside, such as ochre yellow, pink, mint green and sky blue, which support the serene atmosphere of Hans Scharoun's bright and fluid spaces.

In the second half of XX century, many architects dedicated attention to the school environment as an important support for the youngest education intended in a broader sense. As Aldo Rossi said, within the spaces defined by the architecture, the child's imagination has the willingness to build its own space, to add its own personality (Rossi 1977). He argues that children have a different perception of space than adults, just as the use of spaces and objects is different. At the same time the environment induces behavior (Norberg Shulz 1971) and offers many different possibilities, so that the school space cannot be conceived as an empty space to be filled but as a central subject.

In more recent decades many contemporary projects focus on the sustainability and energy efficiency of school buildings but, with reference to the well-being framework for interiors (fig. 1), it can be observed that few pay also attention to a wider concept of well-being.

Among them, the Artemis project at the Melbourne Girls Grammar School (BVN Architects, 2017) is significant as its specific aim is to promote the holistic well-being of both students and staff. In addition to the various functional areas, which are articulated to offer numerous opportunities, the designers were asked to create "Me Zones" for well-being, in different sizes, that are part of the pedagogical model of girl emancipation.

Other projects pay particular attention to aspects that have been recognized above as crucial for well-being: Architects Iredale Pedersen Hook have created the Highgate Primary School (Perth, Australia 2018) trying to promote connection with the environment (with the inclusion of trees, with the use of direct sunlight, with natural materials and colors that represent the different seasons) with creativity (with a varied light, with patterns and colors) and with the reflection of cultural diversity of the student population (with a varied architecture).

Vernacular architecture and tradition lie at the heart of fascinating project of the Indonesia's Alfa Omega school (Banten Province, by Realrich Architecture Workshop 2017): all the materials and local craftsmen were found within five kilometers of the site. So, the children are directly in contact with traditional spaces which bind them to their own cultural roots.

ON THE FIELD EXPERIMENTATION: METHODOLOGY

The theme of well-being, as has been mentioned, acquires different nuances in relation to the cultural context and also in relation to age (Minucciani and Saglar Onay 2019).

For this reason, in order to study well-being in schools, it is very important to decline the concept with specific reference to a pre-adult age, when as explained above the perceptions of spaces are different and when the environment has a strong influence on the development of the person.

Students spend a very important part of their daily life at school, so the school building and its interior spaces really deserve great attention. Therefore, it was intended to investigate the relationship between school spaces and well-being through the direct involvement of the students, as few studies are available in this regard.

In order to proceed with an on-field test, following a homogeneous method, it was essential to target students of different social conditions but in the same age group (11-14). In this sense students of different secondary schools in the Turin area (Istituto Comprensivo Centro Storico of Moncalieri) were asked to answer to a questionnaire that reflects the requirements identified as environmental components of well-being, complying with the theoretical framework of reference (fig. 1): the connection to context, and functional, ergonomic, psychological, social, aesthetic and sensory requirements.

However, more information was needed, in addition to the young people's remarks about the school spaces and furnishings. First of all, it was very important to know the quality standard to which individual students are accustomed: in other words, the type of dwelling they live in undoubtedly influences their perception of spatial quality and their conception of comfort. That is why the questionnaire included some questions about their homes and their habits. As some significant differences could come from gender, also this information was requested. Finally, in order to reflect the cultural differences, within which observations must always be brought back, children were asked to identify the country of origin of their parents.

Although each answer referred in the first instance to a precise well-being requirement, often other requirements were also associated with the question, albeit with less evidence. Furthermore, since students could prioritize different answers within the same topic, it was important to consider the full range of their responses in order to better interpret their preferences. Then a detailed examination of the sample's responses reveals preferences that can be interpreted in relation to the well-being requirements to which they relate, as it will be explained in the following paragraphs.

From the point of view of methodological approach, the first three preferences and the last one were taken into account for the evaluation of the answers, for which a hierarchy of preference was requested. In accordance, the preferences have been scored to report the trends in more detail. The algebraic sums then led to the final score for each answer, which was represented in the diagrams.

Table 1. Structure of the questionnaire

3. STATIC INFORMATION **1.1 PERSONAL INFORMATION** Age / gender / country of origin of parents **3.2 DAILY ROUTINES AND DWELLING** **social life:** How often do you have guests in your house? (everyday, often, almost never) **activity:** what do you do in your free time? (sport, watching television, playing with friends, playing alone on the computer, chatting with mobile phone, other) **your dwelling:** how many rooms (bedrooms, living room, study...) does it have (not included the kitchen and bathrooms)? / How many bathrooms does it have? Does it have a garden? Does it have a garage? **In your bedroom in your room what would you like, and what wouldn't you like?** to be numbered in ascending order from most desired to least desired: to have my own balcony / to have the space to do everything needed / to have comfortable furniture (large desk, comfortable bed, wardrobes wit space for my things etc.) / to have the possibility to decide how to organize it, and what things to put on / to be able to accommodate a friend / to always have a good temperature, and a good light / to have many beautiful and special things that I like (furniture, ornaments, and more) **What is, besides your room, the place in the house where you spend the most time?** Living room, Kitchen, other **What do you usually do there?** Playing alone, playing and spending time with my friends, spending time with my family / other
2. ABOUT SPACES AT SCHOOL **Connection to context** **What is most important to have in the school?** To be numbered in ascending order from most important to least important: Having big windows opening to interesting views / Having a pleasant path to enter / Having a garden or big terraces / Being able to go to exterior space directly from the class **Functional requirements** **What would you like more in the school?** To be numbered in ascending order from most desired to least desired: Having a large gym with all the equipment for the different sports / Having a canteen where the queue is neat / Having lots of bathrooms and sinks so you don't have to queue / Having a large space to do activities with all classes together / Having games and benches in the garden, and sports facilities / Having different labs for each activity (art, technique, science, music) **Ergonomic requirements** **What do you think is most important in the school?** To be numbered in ascending order from most important to least important: having comfortable desks with sockets for charging computers / having seats for spectators in the gym / Having a lunch room where you easily take food, and you see it well / Having in the bathrooms mirrors and sinks at the right height / having enough space in the labs to do your work without being disturbed by others **Psychological requirements** **What is most necessary in the school?** To be numbered in ascending order from most needed to least needed That there are spaces to be quiet alone during the interval / That older and younger children cannot mix (e.g. in bathrooms, canteens, etc.) / That each class has its own space, at its exclusive disposal / That there is a space where everyone can present the things they do (a sport, an artistic ability, some interesting things they do) **Social requirements** **What do you think is most important to have the opportunity to do in the school?** To be numbered in ascending order from most important to least important: Meeting students from other classes / using computers and tablets to play during the interval / studying and doing homework in groups **Sensory requirements** **Which would annoy you most at school?** To be numbered in ascending order from most annoying to least annoying: Being too cold / being too hot / hearing noises from outside or from other classes / having low natural light from the windows and having to turn on the lights **Aesthetic requirements** **What would you like most in school, and what less?** To be numbered in ascending order from most desired to least desired: A lot of colors / many panels, drawings and pictures on the walls / large spaces to play / curved walls / nature in interior spaces

Table 1 details the structure of the questionnaire with the proposed answers. The questions were organized according to three main categories, as shown: static information (about the test subjects and her/his dwelling); spatial preferences (about preferences within each well-being requirement) and hierarchy of well-being requirements (about the grading scale of importance assigned to different well-being requirements).

Although it is possible to extrapolate many interesting reflections from each of the question asked and trends of responses, the most significant selection is reported here. In the following part, the relevant questions will be examined individually, and then concluding reflections will be proposed.

Connection to Context

"Connection to context is the most important design criteria that creates architectural and spatial identity by differentiating buildings/spaces with context related data. Contextual ties can be discussed in terms of location, orientation, entrance, openings, surrounding buildings /spaces, infrastructure and natural aspects. A growing body of knowledge supports the role of contact with nature in human wellbeing" (Saglar Onay, Minucciani 2018).

About connection to context, in the case of school, it relates to the immediate surroundings with which children may have a direct relation. Each response implies a greater or lesser reference to the other requirements, as shown in table 2: that is, given that the main requirement addressed is "connection to context", the different answers call into question some of others, listed in the first column.

In this regard, Figure 2 demonstrates that students' preferences are distributed fairly evenly, but the functional aspects (the direct connection of the classrooms with the outside, the presence of gardens or terraces) seem to be more appreciated than the aesthetic and psychological aspects of a connection with the surrounding environment.

Being able to go to exterior space directly from the class is the first answer to be chosen: this result could be connected to the poor quality of the internal circulation areas (which usually consist of long corridors without appeal, where students crowd uncomfortably in the entrance and exit hours).

Table 2. Connection to context

What's more important at school?	
Further requirements involved	Proposed answers
aesthetic requirement, psychological requirement	having big windows opening to interesting views
aesthetic requirement, sensory requirement, functional requirement	having a pleasant path to enter
functional requirement, social requirement	having a garden or big terraces
functional requirement, psychological requirement	being able to go to exterior space directly from the class

Figure 2. Answers to the question "What is most important to have in school?"

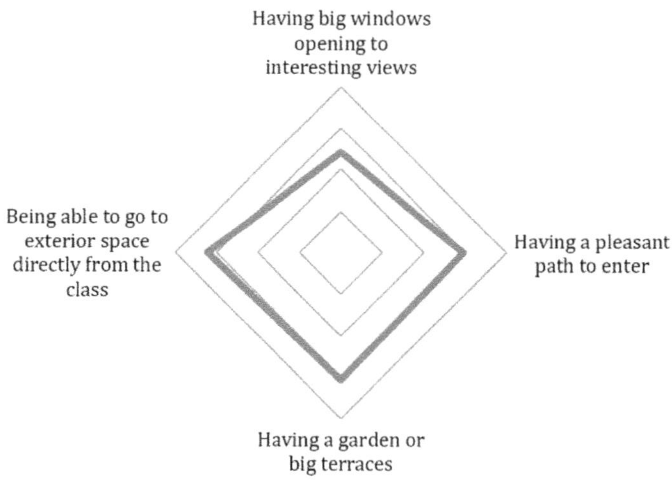

Functional Requirements

In general terms, functional needs can be regarded as the basic needs in environments intended for a certain purpose. The functional requirements may affect the organization of all interior elements such as furniture, lighting etc. The most efficient way of understanding functional requirements is to understand potential user's intended activities and their way of performing these activities. The success in determining these requirements has a significant effect on wellbeing of inhabitants or users (Saglar Onay, Minucciani 2018).

With respect to functional requirements, different types of responses were proposed to young students: responses concerning physical or manual

Table 3. Functional requirements

What would you like more in school?	
different types of responses	Proposed answers
answers concerning physical or manual activities	having a large gym with all the equipment for different sports / Having different labs for each activity (art, technique, science, music)
answers related to a more marked sense of socialization	having a large space to do activities with all classes together / Having games and benches in the garden, and sports facilities
answers referring to functional aspects in a strict sense	having a canteen where the queue is neat / Having lots of bathrooms and sinks so you don't have to queue

Figure 3. Answers to the question "What would you desire more at school?"

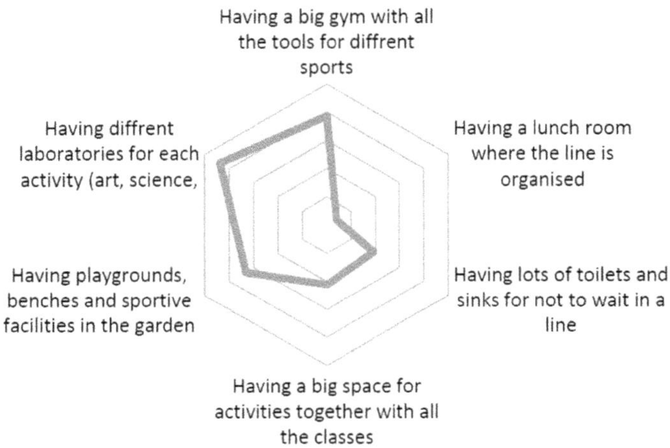

activities, responses related to a more marked sense of socialization and responses referring to functional aspects in a strict sense (see Table 3).

Figure 3 clearly shows, the first class of responses was widely the most popular: within this trend, the possibility of being able to easily carry out different types of manual or artistic activities took priority over activities more related to physical expression such as sport, although much appreciated.

The need to release their physical energy emerges however within the second category of responses, where the recreational aspect and the possibility of staying outdoors were preferred, in a fairly significant way, to activities that involve opportunities for socialization. If the space at their disposal allow them to comfortably carry out these activities, their well-being increases.

Not surprisingly, purely functional aspects relating to certain school facilities or to processes do not seem to be highly voted by students. The most appreciated functional requirements are therefore referred to specific activities that are typical of the educational context.

Ergonomic Requirements

Ergonomic requirements in interiors are related to the connection between human and space, mainly including physical and cognitive aspects, which help to fulfill a certain task (Saglar Onay, Minucciani 2018).

Also in this case the proposed answers involve different aspects of ergonomics: answers concerned with the possibility of carrying out mental or manual activities in a comfortable way, with the necessary equipment and spaces; answers related to socialization needs and finally answers referring to purely ergonomic and organizational aspects (see Table 4).

Table 4. Ergonomic requirements

What do you think is most important in the school?	
answers concerned with the possibility of carrying out their activities in a comfortable way	having comfortable desks with sockets for charging computers / having space enough in the labs to do your work without being disturbed by others
answers related to socialization needs	having seats for spectators in the gym
answers referring to purely ergonomic and organizational aspects	having a lunch room where you easily take food, and you see it well / having in the bathrooms mirrors and sinks at the right height

Figure 4 confirms the trends that have already emerged in the case of functional requirements, to which ergonomic requirements are closely

Figure 4. Answers to the question "What do you think is most important in the school?"

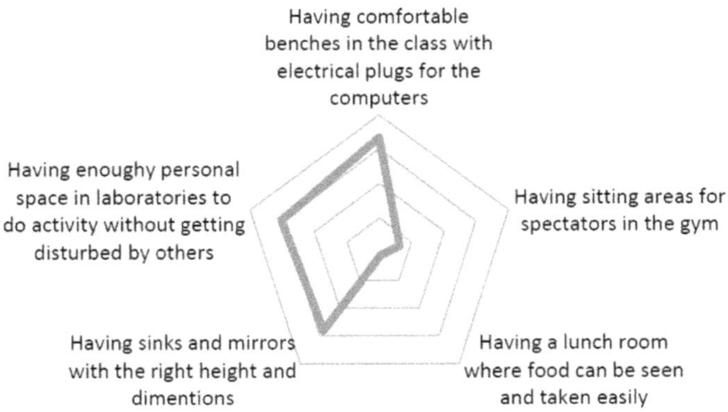

connected: even in this case, in fact, students have clearly expressed the desire to be able to carry out activities (not only physical and manual, but also mental) in suitable and comfortable conditions; indeed, the needs related to mental learning (the comfort of the desks, the availability of computers) are even more relevant than those related to physical and manual activities. Very little importance was given to socialization aspects, in contrast to other trends that emerged in the questionnaire, can be explained by the focus on activities in themselves.

Conversely, the answers concerning purely ergonomic and organizational aspects give results that seem to be in contrast with each other: in the case of bathrooms there is a great demand for suitable equipment according to the needs of the children, while the lunch room facilities (as already in the case of functional requirements) don't seem particularly important for the students. We can suppose that where the activity (in the first case, using bathrooms) is strictly functional and has no other implications, students concentrate on it and inevitably perceive the greater or lesser comfort of the equipment and spaces; but if the activity (in the second case, eating together with other students) has strong recreational and social implications, its strictly functional and ergonomic needs are decidedly overshadowed because children - unlike adults - seem not to perceive them and to adapt themselves very naturally to less comfortable situations.

Psychological Requirements

Self-determination theorists (Ryan & Deci 1985 and 2000) maintain that well-being hinges on the fulfillment of three basic innate psychological needs which are defined as competence, relatedness and autonomy, for which privacy (strongly linked to space) is indispensable.

Table 5. Psychological requirements

What is most necessary in the school?	
privacy, security	that there are spaces to be quiet alone during the interval / That each class has its own space, at its exclusive disposal
security, privacy	that older and younger children cannot mix (e.g. in bathrooms, lunch rooms, etc.)
social visibility	that there is a space where everyone can present the things they do (a sport, an artistic ability, some interesting things they do)

Psychological aspects, intended as the psychological effects that an architectural environment can generate on users, in the case of school spaces are very complex and certainly very different from those of adults. Psychological requirements have been here intended as privacy, security and the desire to acquire "social visibility" (i.e. a communication requirement). Some answers have implications from more than one point of view, as shown in the table 5.

Figure 5. Answers to the question "What is most necessary in the school?"

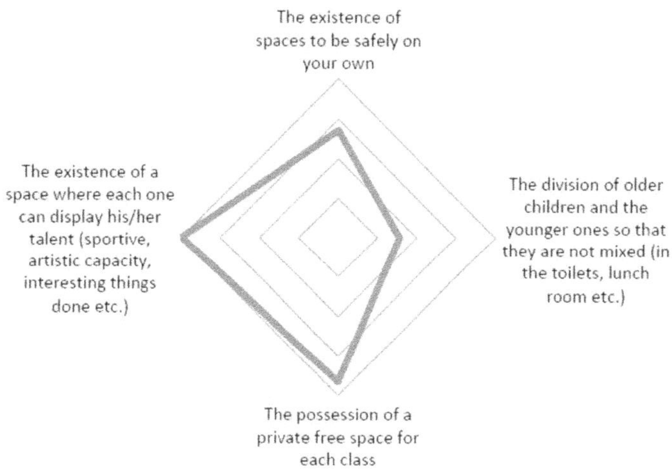

Figure 5 shows very clear tendencies. Over all the others, the desire to acquire "social visibility" emerges. Both the comparison with the other students and (above all) the need to know/be known seems to demonstrate a great need for communication, that is a need of social nature.

At the same time, the opposite need (having private spaces in order to control and even avoid interactions with other students) also emerges as very important.

In particular, investigating this finding in relation to gender and country of origin, the demand for spaces where you can be alone without being disturbed seems to be stronger in males than in females (for whom it is still important) and especially for children of non-Italian origin: this could easily be interpreted as the result of a process of integration not yet complete and satisfactory.

Table 6. Social requirements

What is most necessary in the school?	
social (socialization)	meeting students from other classes
social (isolation), functional	using computers and tablets to play during the interval
social (socialization), functional	studying and doing homework in groups

Social Requirements

Well-being of individuals is related to wellbeing of others and the interaction between individuals and groups. Although it may seem independent of spaces, our activities promoting social wellbeing are directly affected by spatial properties that enrich social life. In this sense, interiors should provide the chance for social encounters at desired level for everyone. Moreover, they should support and encourage social activities creating a balance between public, semi-public and private places (Saglar Onay, Minucciani 2018).

As it emerges from what was previously discussed, social aspects are quite pervasive in the school environment, where collective aspects are very important. This question, which complements the above considerations, aims to better understand how important these aspects (compared to individual activities) are.

Social issues can be divided into two macro-categories: the first one is connected to the need to socialize (as a goal in itself, to build relationships of knowledge and friendship, or simply for leisure and recreation), the second one is connected to the need to be isolated from the community in which you are constantly immersed: it does not necessarily coincide with the need for privacy, which, as stated above, is mainly of a psychological nature. Furthermore, the different meanings of the answers can be related to other well-being requirements, primarily to the functional aspect, as shown in table 6.

Figure 6 shows, perhaps unexpectedly, that socialization in itself is not considered particularly important, while socialization related to learning and work collects the largest amount of preferences.

The tendency to be isolated is not much less (although with the specific clarification that such a private time can be spent in using computers or tablets).

These answers also provide interesting suggestions for the definition of school environments, that must be flexible in order to meet both socializing and isolating needs, but at the same time it cannot be an indistinct place.

Figure 6. Answers to the question "What is more important to be able to do at school?"

Sensory Requirements

Sensory requirements are extremely effective on well-being, intended in its comfort component. Ventilation, lighting, heating, cooling and acoustics are issues to be handled with special expertise – but sensory requirements include further components (Saglar Onay, Minucciani 2018). Pallasma argues that qualities of space, matter and scale are measured equally by the eye, ear, nose,

Table 7. Sensory requirements

Which would annoy you most at school?	
sensory	being too cold / being too hot
sensory, functional, psychological	hearing noises from outside or from other classes
sensory, psychological, connection to context	having low natural light from the windows

skin, tongue, skeleton and muscle (Pallasmaa 2005). Relatedly Zumthor states that we perceive atmosphere through our emotional sensibility, which is a form of perception that works incredibly quickly (Zumthor 2006).

The main sensory stimuli have been proposed, and also in this case they may be related to other aspects of well-being (See Table 7).

Some of them have an exclusively sensory meaning, able to significantly alter the perception of physical well-being. Others, although related to situations of comfort or discomfort of a physical nature, have nonetheless

Figure 7. Answers to the question "Which would annoy you most at school?"

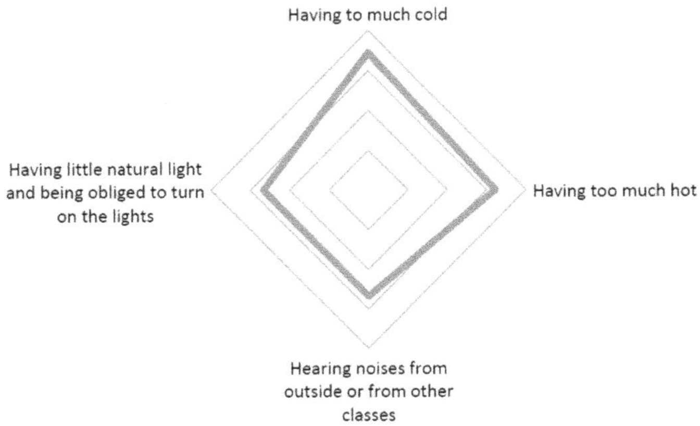

psychological implications (first of all) but are also in connection with other well-being requirements.

The distribution of the answers, as shown in Figure 7, is quite balanced but shows a fairly evident shift in the "hub": in fact, the conditions that have an exclusive effect on physical comfort have gathered the greatest preferences, with a small but significant difference for the answer "being too cold", which therefore seems to be a more negative situation than "being too hot". Secondly, the sensory situations that hinder the performance of the function (i.e. study) causing difficulties in concentration have also been reported frequently, while the least problematic situation among all is the one in which natural light is scarce, leading to the use of artificial light. The result could be a confirmation of the fact that the psychological aspect of connection to context is not as much felt by the children as the aspect related to leisure or play.

Aesthetic Requirements

Although referring to aesthetics immediately concerns the beautiful shaping of a building, this requirement is hard to be defined. The founder of aesthetics, German philosopher Alexander Gottlieb Baumgarten introduced a "gnoseological subjectivism", then the ideal of beauty as something that exists independently from cognitive experience, disappears. At the turn from of the XIX century, Georg Wilhelm Friedrich Hegel changes the focus of aesthetic considerations from the form to the meaning (Stevanović 2013).

"Although aesthetics has been a major concern in architecture, there is thus far no models or theories specifically dedicated to the psychology of architectural aesthetics (…) the empirical studies of architectural aesthetics have only just begun to gain more attention in experimental psychology and neuroscience" (Ho 2020). Referring to perception of architectural spaces, indoor or outdoor, it can be understood in terms of scene perception, and it should be noted that in youngest people the aesthetic judgement is different and less influenced by cultural meanings than in adults.

From a methodological point of view, this is thus a question whose possible answers have been elaborated with a greater degree of uncertainty. In general, different aspects of aesthetic requirements have been considered: the decorative effect (which provides a series of stimuli, above all sensory and visual), the bond with functional and/or social repercussions, psychological relax, the connection with nature and the formal and symbolic value of organic shapes (as shown in table 7).

To help interpret the question about aesthetic requirements, different answers were accompanied by exemplifying images (this is the only case where images were used). Aesthetic judgement, cannot disregard the physical "immersion" in space, since it is influenced by atmosphere that can be appreciated only if one is really in that space. Conversely, an image relies on only one sense (sight) while the perception and therefore also the aesthetic effect of a space is given by the combination of all the senses, even if with a prevalence of sight. Besides, while the images could facilitate the students' choices, at the same time their impact could prevail over the real meaning of the questions.

Table 7 shows the images proposed for each answer choice. The first three responses represent cases where spaces quite clearly reflect also functional and social requirements. The last answer contemplates a sort of connection with the context, while the fourth answer is the only one that apparently has an exclusively aesthetic character, but can easily imply a reference to psychological requirements (the shape and the organic line can be more welcoming and more reassuring).

The result, as presented in Figure 8, shows clear preferences. Unexpectedly, the presence of many colors does not seem to reap much success despite it is from many designers felt that many colors are suitable for spaces for young students.

An equally low level of preference was obtained from the second answer, which carries with it both social and functional connotations (linked to school activities), while the third answer gets a little greater success, having also social and functional implications (but linked to play and recreation).

Figure 8. Aesthetic requirements

What would you like most in school, and what less?

concerning the sensory effect of decoration		a lot of colors
relating to socialization needs, and functional requirements		many panels, drawings and pictures on the walls
referring to functional aspect (leisure), and social life at school		large spaces to play
relating to aesthetic value and effect of organic shapes (less usual, and more comfortable)		curved walls
concerning the fusion nature-architecture, and the connection to context		nature in interior spaces

The fourth answer collected many preferences, but its absolutely dominant color (pink) could have influenced, for cultural reasons, the choices of males (even if many have indicated it anyway). The preference for simple, smooth and curved contours on judgements of aesthetic pleasure date from the earliest work in empirical aesthetics (Stratton, 1902; Valentine, 1913), and more recent studies confirmed this predilection in subjects across different age groups (Quinn et al 1997; Jadva et al 2010) and cultures (Gómez-Puerto et al 2018). Intimacy and ceiling height seems to be other interior aspects affecting aesthetic judgement (Ho 2020).

Figure 9. Answers to the question "What would you like most in school, and what less?"

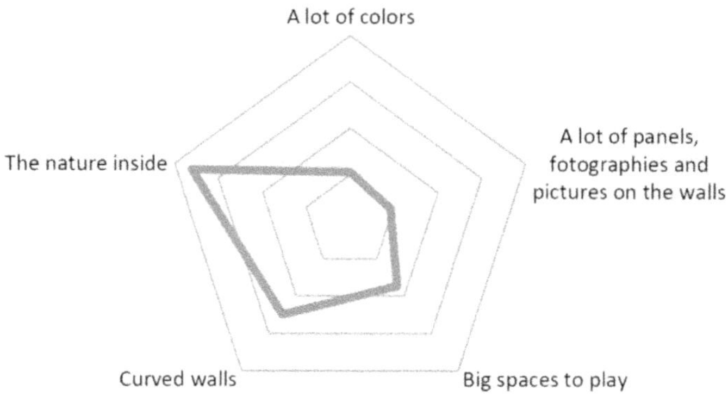

As can be seen from Figure 9, the presence of natural elements within the school environment gathered the greatest number of preferences. Not only the connection with context, but also a greater closeness and union with nature is a desire that the students have expressed and that is consistent with their soul, because young and very young people feel a greater affinity with nature than adults.

Table 8. Hierarchy of requirements for well-being, positive question

"In which of these cases do you feel (or would you feel) happier at school?"	
connection to context	when around me there is nature
functional requirements	when I have the space and tools to do my favorite activities
ergonomic requirements	when things around me make me feel comfortable
psychological requirements	when I can be at peace without being disturbed in the interval
social requirements	when I can meet a lot of friends and chat with them
sensory requirements	when I'm comfortable, neither too hot nor too cold
aesthetic requirements	when there are beautiful furniture and colors and objects that I like around me

Hierarchy of Requirements for Well-Being

The two final questions relate to hierarchy among all well-being requirements.

Figure 10. Answers to the question "In which of these cases do you feel (or would you feel) happier at school?"

It was conducted through a crosscheck, first asking the test subjects in which situation they felt more happy, and then in which situation they felt sad. The results do not always coincide. Table 8 shows the different answers proposed to the students, being representative of each requirement.

The diagram (Figure 10) shows strong differences between the different answers. It clearly illustrates that the functional requirements are considered more important than any other. Children demonstrate a great need to experiment and work in different ways (not only about intellectual work, but also physical and manual ones) and consider the availability of the necessary space for these activities as the most important requirement.

Table 9. Hierarchy of requirements for well-being, negative question

In which of these cases do you feel (or would you feel) sadder at school?	
connection to context	in a classroom without windows, or with curtains always lowered
functional requirements	when there is no space to do the activities I like
ergonomic requirements	when desks, chairs and other things are uncomfortable
psychological requirements	when I don't feel safe (older kids might bother me, or something else)
social requirements	when there's no space to stay with friends
sensory requirements	when I am too hot, or too cold, or there is too low light
aesthetic requirements	when walls are dirty and damaged, and desks and chairs and lockers are ugly

Predictably, also social requirements are considered very important, while at the other extreme psychological requirements appear very underestimated, and even aesthetic requirements are not felt to be very important. The connection to the context is perceived as important if it means "natural" context. The ergonomic requirements have received a considerable number of preferences too, although children may seem do not pay too much attention to it, and

Figure 11. Answers to the question "In which of these cases do you feel (or would you feel) sadder at school?"

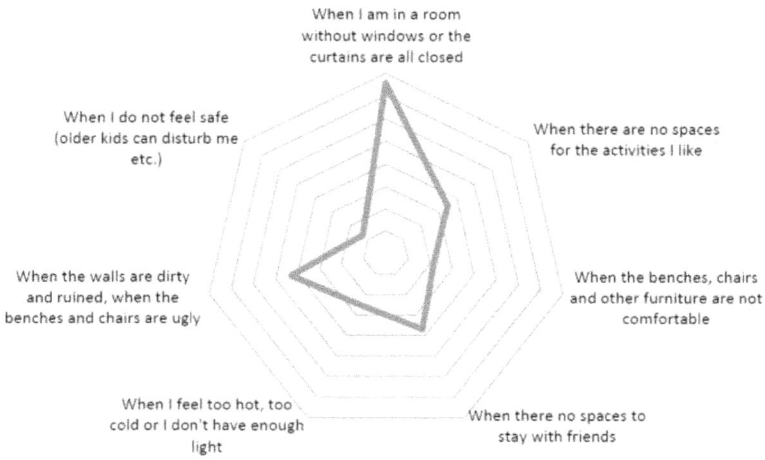

sensory requirements are also taken into account.

Crosschecking was made possible by the last question, which is detailed in Table 9.

The question on how important the lack of certain requirements is, turns the situation upside down quite clearly. The lack of connection to the context is considered the most negative, and an aesthetically degraded situation is in second place (Figure 11).

Social requirements have been assessed in third place and functional requirements only in fourth place. Sensory and ergonomic requirements are assessed even less, and ultimately psychological ones.

Combining these two diagrams it emerges that connection to context is the most important requirement, followed by the social and functional requirements.

Influences on Preferences: Gender and Economic Condition

The observation of the data also suggests some further considerations about gender, since often boys responded differently from girls. Examining in detail the expressed answers and preferences, in some cases they are exactly

Figure 12. Answers to the question "In which of these cases do you feel (or would you feel) happier at school?"

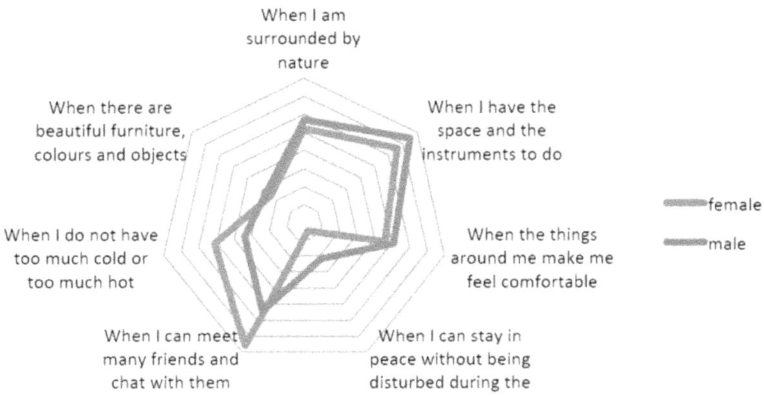

the same, in others they are exactly opposite and overturn the hierarchy: for example, with regard to social requirements, the possibility of using PCs and tablets in the break time is considered the least important by girls and the most important by boys.

These differences / coincidences clearly emerge in Figure 12, which simultaneously shows the hierarchy of preferences for males and females in comparison.

In particular, in the question "In which of these cases do you feel (or would you feel) happier at school?" some answers are very similar, while in the psychological and social requirements we can observe the major differences. Females show a much more marked tendency to appreciate environments that facilitate relationships with friends and socializing activities, and they pay more attention to sensory well-being related to a comfortable temperature. Males show a greater sensitivity to the psychological requirement of privacy, which is linked to a need for security but indicates a lower propensity to socialization than females.

Crosschecking allows to better clarify preferences and hierarchies of responses. By answering the symmetrical question "In which of these cases do you feel (or would you feel) sadder at school?", girls focus their preferences on three requirements: connection to the context, social requirements and aesthetic requirements. Males have less pronounced inequalities between responses, which however also indicate their preference for connection to context and aesthetic requirements: however, they definitely underestimate the social requirements in favor of functional ones. They also attribute much more importance to ergonomic requirements than females (Figure 13).

Figure 13. Answers to the question "In which of these cases do you feel (or would you feel) sadder at school?"

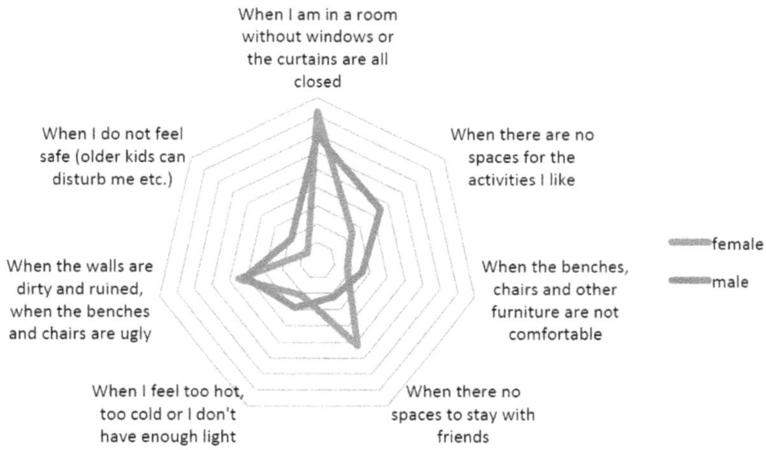

The collected data allow us to consider the social status of the children, which can easily be deduced from the description they provide about their home: it is a non-exhaustive indicator and absolutely it would not be considered entirely correct, but in our case (since we are talking about places where they inhabit and live) the conditions in which a student is used to live affect the perception she/he has of the school environment, in comparison. There are in fact students who live in very large houses, with gardens and many services, others who are forced to sleep in the kitchen because they do not have their own bedroom.

From this point of view, the sample examined is very representative for two reasons: first, because the two schools surveyed have different type of users because of the mixed and varied character of the quarter (Moncalieri) that includes very popular and modest areas as well as very prestigious and exclusive ones, such as the less populated hill and some parts of the historic center that has recently been restored and enhanced.

The results are ambivalent. In general, the economic conditions do not seem to generate very divergent responses, apart from the fact that children who enjoy less comfort are more sensitive to them, while those who are used to have them, take them for granted and probably underestimate them: the most significant example is the availability of a garden, that is the connection with the natural context.

FUTURE RESEARCH DIRECTIONS

The questionnaire collected many data that can still be interpreted according to different perspectives: first of all, the differences between the sexes, or the social and cultural conditions, can be further investigated.

The issue of well-being in the school environment is complex and very important in relation to long-term educational effects. There is no rich literature on the subject but some studies have been carried out. Lester and Cross's study focuses on different components of well-being: safety at school, connectedness to teachers, connectedness to school, and peer support. It also distinguishes emotional well-being and mental well-being (Lester and Cross 2015). However, much work still needs to be done in this regard.

It would be very useful to carry out this survey also on other age groups, younger children (6-10) and older children (15-18), to observe how the concept of well-being at school evolves. On the other hand, in school it would also be necessary to do the same survey on teachers, for whom school is the place of work but certainly, due to its peculiar nature, it cannot be assimilated to office spaces, where there is already a lot of literature on well-being. Students' well-being cannot disregard the teachers well-being, then it can be said that the well-being in school buildings is, more than in other workplaces, given by the average state of satisfaction or dissatisfaction of everyone, teachers and learners. The school is like a restricted community, where social aspects are very important (relations between teachers and between students, student-teacher relations - to which other actors such as parents are added). Moreover, the teachers observe the children and can interpret with great lucidity the requirements for their well-being and for their best school performance. Such a study has yet to be carried out in the field, and has the potential to lead to interesting results.

CONCLUSION

This study, focusing on well-being in school environment based on the "Well-being Framework for Interiors", provided several elements to be deepened.

The caution taken by the students in expressing their preferences and hierarchies in the answers shows how, even in a very particular age such as pre-adolescence, the interior environment in which they spend most of their time is significantly important, and how not only boys and girls pay attention to it but they also have precise ideas on how much it makes their lives more comfortable and pleasant.

First of all, the framework has proven to be applicable to very young age groups. Their answer confirm that the perception of well-being is the result of a complex series of factors, in which reactions of a physical and physiological nature merge inextricably with cultural, social, psychological, even economic factors – then the concept of well-being cannot be generalized.

Furthermore, in the school environment it is evident that "the concept of wellbeing comprises two main elements: feeling good and functioning well" (Jackson 2013): and these two aspects are actually present to different degrees in all of the questions that have been asked to the students, and they generally seem to consider the second aspect as more important.

As Jackson points out, well-being has not clear opposite, and both well-being and "unwell-being" represent an ideal, a generically desirable or undesirable state. This could explain some apparent contradictions between the students' answers: imagining certain conditions prefigured by the answers, the students reacted aiming to maximize well-being, and aiming to minimize "unwell-being".

Finally, the above contradictions can also be explained by the fact that well-being is a holistic concept, for this reason it is complex, multi-faceted and represents a balance between different tensions. Being a social and cultural construct, as well as a mental one, well-being must be interpreted more than established and, above all, it cannot be guaranteed by universal guidelines.

The scarcity of specific studies concerning well-being at school (with specifics attention to very limited age groups) is certainly a gap that must be filled.

REFERENCES

Amabile, T. M., & Kramer, S. J. (2011). *The progress principle: Using small wins to ignite joy, engagement, and creativity at work*. Harvard Business Review Press.

Burke, C. (2005). Containing the school child: Architectures and pedagogies. *Paedagogica Historica*, *41*(4-5), 489–494. doi:10.1080/00309230500165635

Gómez-Puerto, Rosselló, J., Corradi, G., Acedo-Carmona, C., Munar, E., & Nadal, M. (2018). Preference for curved contours across cultures. *Psychology of Aesthetics, Creativity, and the Arts*, *12*(4), 432–439. doi:10.1037/aca0000135

Gur, E. (2014). The effect of physical and environmental factors of a 'child development center' on a center's selection. *Archnet-IJAR*, *8*(3), 136–148. doi:10.26687/archnet-ijar.v8i3.455

Hille, R. T. (2011). *Modern schools. A century of design for education*. John Wiley & Sons.

Ho, P. K. (2020). *The psychological basis of aesthetic preferences: Insights from facial attractiveness and architectural appreciation. Master of Philosophy*. Trinity College Institute of Neuroscience, University of Dublin.

Huppert, F. (2008). *Psychological well-being: Evidence regarding its causes and its consequences*. Foresight Mental Capital and Wellbeing Project.

Jackson, N. J. (2013). Exploring subjective wellbeing and relationships to lifewide education, learning and personal development. Lifewide Learning, Education & Personal Development e-book.

Jadva, V., Hines, M., & Golombok, S. (2010). Infants' preferences for toys, colors, and shapes: Sex differences and similarities. *Archives of Sexual Behavior*, *39*(6), 1261–1273. doi:10.100710508-010-9618-z PMID:20232129

Kozulin, A. (2003). *Vygotsky's Educational Theory in Cultural Context*. Cambridge University Press. doi:10.1017/CBO9780511840975

Lester, L., & Cross, D. (2015). The relationship between school climate and mental and emotional wellbeing over the transition from primary to secondary school. *Psychology of Well-Being*, *5*(1), 9. doi:10.118613612-015-0037-8 PMID:26516619

Merleau Ponty, M. (1945). *Phénoménologie de la perception*. Gallimard.

Minucciani, V., & Saglar Onay, N. (2019). Well-being age and space. *Journal of Urban Design*, *25*(2), 275–286. doi:10.1080/13574809.2019.1667758

Norberg Shulz, C. (1971). *Existence, space and architecture*. Praeger Publishers.

Nussbaum, M. C. (2000). *Women and human development: The capabilities approach*. Cambridge University Press. doi:10.1017/CBO9780511841286

Olds, A. R. (1979). Designing developmentally optimal classrooms for children with special needs. In S. J. Meisels (Ed.), *Special Education and Development: Perspectives on Young Children with Special Needs* (pp. 91–138). University Park Press.

Pallasmaa, J. (2005). *The eyes of the skin*. John Wiley &Sons.

Piaget, J. (1980). Adaptation and intelligence: Organic selection and phenocopy. Chicago: University of Chicago Press. (Original work published 1974)

Piaget, J. (2001). The psychology of intelligence. New York: Routledge. (Original work published 1963)

Piaget, J., & Inhelder, B. (1967). *The child's conception of space*. Routledge.

Quinn, P. C., Brown, C. R., & Streppa, M. L. (1997). Perceptual organization of complex visual configurations by young infants. *Infant Behavior and Development*, *20*(1), 35–46. doi:10.1016/S0163-6383(97)90059-X

Rossi, A. (1977). Scuola elementare a Fagnano Olona. *Lotus International, 15*.

Saglar Onay, N., & Minucciani, V. (2018). Wellbeing Framework as a Contributor to Sustainability. In *WIT Transactions on Ecology and the Environment, 217* (pp. 699–710). WIT Press.

Seligman, M. E. P. (2002). *Authentic happiness: Using the new positive psychology to realize your potential for lasting fulfillment*. Free Press.

Stevanović, V. (2013). Ideological assumptions in aesthetic judgment of architecture. *SPATIUM International Review*, *30*(30), 40–46. doi:10.2298/SPAT1330040S

Stratton, G. M. (1902). Eye-movement and the aesthetics of visual form. *Philosophische Studien*, *20*, 336–359.

Türel, A., & Gür, A. (2019). Effects of primary school's physical environment on children's spatial perception and behavior The case of Kagithane, Istanbul, Turkey. *Archnet-IJAR: International Journal of Architectural Research*, *13*(2), 425–443. doi:10.1108/ARCH-12-2018-0048

Valentine, C. W. (1913). The beauty of form. In An introduction to experimental psychology of beauty. New York: Dodge.

Vygotsky, L. S. (1986). *Thought and language (Newly revised and edited by Alex Kozulin)*. The Massachusetts Institute of Technology, MIT Press.

White, J. (2011). *A guide to make children's lives more fulfilling*. Routledge.

Yavuzer, H. (2000). *Çocuk psikolojisi [Child Psychology]*. Remzi Kitabevi.

Zumthor, P. (2006). *Atmospheres*. Birkhauser.

Chapter 9
Cultural Components of Well-Being and Vernacular Examples

ABSTRACT

Vernacular architecture expresses a highly sustainable approach to living as a result of strong connection to cultural and environmental context. On the other hand, today's architecture attempts to create artificial solutions that consume the environment and lack the authenticity of vernacular solutions. In this sense, vernacular living environments need to be conserved, appreciated, and studied carefully because of their great potential to inspire the new generation of living environments. Previous research on the topic has mostly focused on the use of local materials and inherited construction techniques, which are extremely beneficial. Starting from a short survey on international declarations and charters, and discussing the strong connectedness between vernacular design principles and human well-being as an outcome, this chapter will lead the discussion deeper on interiors. Interiors from selected cultural contexts will be examined in order to attract attention to the cultural components of well-being and their meaningful reflections on the creation of interior space.

DOI: 10.4018/978-1-7998-4231-6.ch009

INTRODUCTION

Conservation should not be evaluated only in physical terms but also as an important source of intangible knowledge and inspiration for future generations. In this perspective the conservation of interiors is extremely important as they are strongly connected to everyday life representing all cultural and social aspects related to living. On the other hand, because of this strong connection to life, interiors are the most ephemeral components of our living environments changing related to shifting needs and ways of living of the society. Thus it becomes extremely difficult and even constraining to conserve interiors as they are. At this point, conservation of the approach (of creation of interior space) and its specific message becomes more important and meaningful than physical conservation.

In this chapter well-being in interiors will be discussed through vernacular examples. For this reason the concepts of well-being in interiors and vernacular design principles will be structured in order to recognize the design principles in the intersection of these two important concepts. Well-being in interiors will be structured using the framework discussed in Chapter 5. The well-being framework for interiors is composed of several requirements that built up subjective well-being (Figure 1). The structure of the framework is based on contextual, functional, psychological, social, sensory, aesthetic and ergonomic requirements, which are always in relation to each other. These requirements are different than general requirements of design for well-being as they are strongly connected to space and context. Moreover all these requirements are in relation to each other and they are subject to cultural differences regarding users with different backgrounds. Connection to context is the key element of the well-being framework. All other requirements have contextual dimensions and they are always in relation to contextual ties. So, connection to context is the most important design input that creates architectural and spatial identity by differentiating buildings/spaces with context related data. Functional requirements are concerned with creating spatial solutions that support human activities. Ergonomic requirements in interiors are related to the connection between human and space, mainly including physical and cognitive aspects, which help to fulfil a certain task. Psychological requirements are more indirect as they deal with the outcomes of the interaction between human and space. Social requirements deal with the spatial opportunities for the desired level of interaction between people. Sensory requirements are concerned with both quantitative factors like heating, illumination, acoustics, sound control

Figure 1. Well-being Framework for Interiors

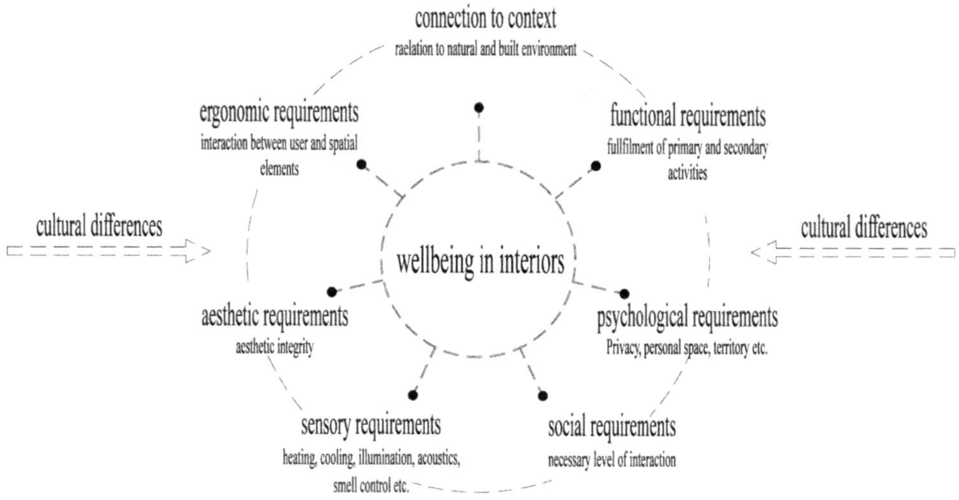

etc. and their reflections on human senses. And lastly, aesthetic requirements are also important as they refer to the integrity and harmony of all spatial elements that can be pleasing to human senses and can promote well-being.

Regarding vernacular architecture, one of the most important international documents is The Charter on the Built Vernacular Heritage (ICOMOS, 1999). The document states that vernacular buildings are the fundamental expression of the culture of a community, of its relationship with its territory and, at the same time, the expression of the world's cultural diversity. According to the charter, vernacular buildings can be recognised by

- a manner of building shared by the community;
- a recognizable local or regional character responsive to the environment;
- coherence of style, form and appearance, or the use of traditionally established building types;
- traditional expertise in design and construction which is transmitted informally;
- an effective response to functional, social and environmental constraints and
- an effective application of traditional construction systems and crafts.

Although the Charter's main focus is on the conservation of the architectural structures of vernacular buildings, it also mentions intangible aspects and issues like change, adaptation and integrity that are aspects connected to the interior environments of vernacular heritage. It underlines that vernacular architecture embraces not only the physical form and fabric of buildings, structures and spaces, but the ways in which they are used and understood, and the traditions and the intangible associations, which attach to them. It states that changes over time should be appreciated and understood as important aspects of vernacular architecture. Conformity of all parts of a building to a single period, will not normally be the goal of work on vernacular structures.

Regarding ways of living of public, vernacular architecture holds the most important clues about interior aspects conveying living styles and attitude towards natural environment and resources. In this sense this chapter focuses on vernacular architecture as a true expression of everyday life, of great part of public, more diffused. In particular vernacular architecture communicates the coeval concept of well-being, as it is not detached from daily life, environment and cultural values.

Observing conservation of interiors we can talk about two different prospects, which are physical conservation and the appreciation of the approach, skills and attitudes towards the natural and built environment. As stated above preservation of interiors especially in vernacular environments is not very easy because of the great changes in society and ways of living. On the other hand, few conserved examples give very important clues about their concept of sustainability and well-being and are really worth safeguarding.

Nowadays, well-being is considered as depending on solutions globally valid (see for instance WELL certification) while vernacular architectures differ regarding all requirements related to well-being, that are influenced by cultural differences. On the other hand the effects of globalization are very threatening on the surviving vernacular environment. Due to the homogenisation of culture and of global socio-economic transformation, vernacular structures all around the world are extremely vulnerable, facing serious problems of obsolescence, internal equilibrium and integration (ICOMOS, 1999).

INTERIORS OF VERNACULAR ARCHITECTURE

Among all well-being determinants, the physical environment can be regarded as the most important concrete aspect that can be controlled by the designer and can directly affect human well-being. Regarding the environment as a whole, this work focuses on interiors as the basic spatial elements where human perform great part of their basic activities. In this part of the study, vernacular homes from different cultural contexts will be evaluated regarding the relation between well-being requirements stated in the well-being framework for interiors (Figure 1) and the vernacular design principles obtained by assessing the Charter on the Built Vernacular Heritage (ICOMOS, 1999).

- Responsive to environment
- Traditional expertise
- Relation to social structure
- Relation to everyday life
- Coherence of style

Connection to Context

Regional character responsive to the environment and strong relation to context is the key aspect of all vernacular solutions as they aim to benefit from the existing environment as much as possible. The evaluation of existing aspects includes local materials, climate, knowledge accumulation and other indigenous sources. In this sense, sustainability becomes a natural outcome of this connectedness between people, environment and sources.

Vernacular buildings have always taken great benefit of local aspects also helping them to acquire a unique design language. Even simplest costless materials such as earth, straw and sand have been used to obtain composite materials. Potential of these materials were discovered after hundreds of years of expertise and they were processed effectively in order to answer to all needs and constrains. For the Japanese house the most important contextual aspect is the strong relation between inside and outside characterised by the Japanese garden. For the African teleukaka the construction method has evolved together with an ingenious material, mixture of soil, thatch, grasses and animal dung effectively responding to environmental aspects and climate. For the Malay house the response to climate, natural ventilation inside the house shapes the overall spatial design. For the Turkish house, the timber frame construction system can be considered as a local construction system taking the advantage of local materials and answering effectively to the danger of earthquakes as a natural phenomenon common in the region. Considering examples from different parts of the world, it can be seen that the concept of space changes greatly while the main principles remain the same. These clever solutions created by local people, contribute greatly also to the concept of well-being.

Table 1. The relation between well-being requirements in interiors and vernacular design principles

		Vernacular Design Principles				
		Responsive to Environment	**Traditional Expertise**	**Relation to Social Structure**	**Relation to Everyday Life**	**Coherence of Style**
Well-being requirements in interior space	**Contextual**	Local materials, climate changes	Local construction techniques	Effects on spatial organization	Relation between interior and exterior	Context related data
	Social	Relation between interior and exterior	Involvement of community in the construction	Concurrency with family structure	Concurrency with social rituals	Social related data
	Psychological			Fulfillment of autonomy, relatedness to social environment	Fulfillment of the need of competence	
	Functional			Concurrency with social activities	Concurrency with everyday activities, agricultural work etc.	Function related data
	Ergonomic	Use of local materials	Use of traditional objects and furniture for rituals and activities		Spatial elements and furniture suitable for activities	Human related data
	Sensory	Response to climate	Use of natural materials and craft		Sensorial response activities	Sensory related data
	Aesthetic	Architectural features for specific climate and envi. like roofs, openings etc.	Continuation of collective memory	Maintenance of sense of community		Repetitive architectural features, details, ornaments
	Cultural	Context related Identity	Passing cultural values to younger generations	Appreciation of cultural values, ex: respect for elderly	Respect to rituals and relation to habits	**Unique identity**

Functional Requirements

In vernacular homes, space is strongly related to everyday life and activities necessitating consideration of functional requirements. For a certain interior, functional requirements may effect the organization of all interior elements such as furniture, lighting etc. The most efficient way of understanding functional requirements is to understand potential user's intended activities and their way of performing these activities. For multi purpose spaces, functional requirements need to be handled with a flexible approach, which

Figure 2. Vernaular interiors from different cultures, orderly Italian cascina (Santomauro, 2014), Japanese minka (Kawashima, 1986), igloo (www.pinterest.com), Turkish Safranbolu house (Küçükerman, 1989), traditional Malay house (Hidayahtuljamilah, 2012) and African teleukaka (Nelson, 2007).

give the possibility to use the same space for different purposes at different times or even for different user profiles (Saglar Onay, Minucciani, 2018).

The typical, north Italian Rural House (the so called Cascina), is a complex featured by the perfect coexistence between men, animals and agricultural work. It is usually articulated around a central courtyard, which is the hub of all activities and the heart of everyday life: here the products converge and there are appropriate spaces for their preservation (ventilated areas under the roofs). Small animals roam freely in these central areas, equipped with spaces for their night shelter. On the ground floor, porticoed spaces allow the continuation of work in different weather conditions.

The Cascina hosts a small, relatively self-sufficient community, where roles and spaces are clearly defined. One wing of the complex is used as a dwelling: on the ground floor there is a large kitchen with fireplace, which is the meeting place of the family and where most of the domestic activities are carried out. This room has a temperature always comfortable and can heat the upper floor, where there are the bedrooms. The furnishings are always very robust, simple and above all practical. In the Cascina well-being is given

Figure 3. Italian Cascina featured by the perfect coexistence between men, animals and agricultural work

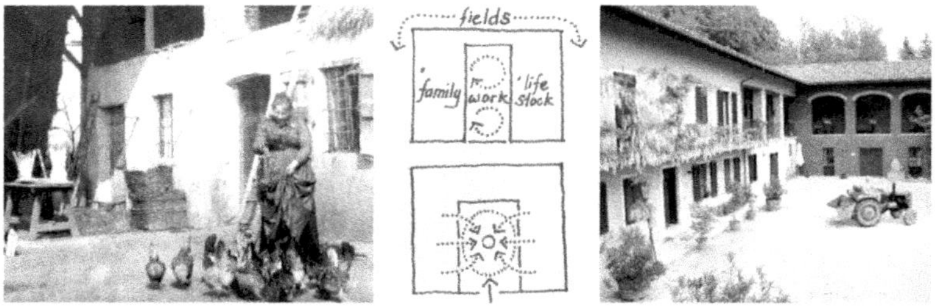

by a balanced relationship with nature, work and environment (Saglar Onay, Minucciani, 2018).

Ergonomic Requirements

As vernacular structures are built by traditional expertise based on knowledge transmitted informally by the members of certain community, design principles are highly affected by the ways of using space. Main activity sets and their relations with space lead to a strong understanding of ergonomic space. Traditional craft products such as hand tools may evolve over a period of time by a cyclic process of trial and error, which is called vernacular ergonomics (Pheasant, 1991). For home environments vernacular ergonomics is a natural outcome of generations of continuous feedback of living experiences. In this sense we can note a direct correspondence between spatial features like furniture, accessories, tools etc. and the main activities (eating, sitting, cooking, sleeping etc.) that take place in home environments. Thus, the design features of interior components follow the ways the activities are fulfilled including both physical and cognitive aspects.

The interior environment of the traditional Turkish house reflects the ways of living in the whole spatial layout as well as in furniture scale. In the typical room of the Turkish House necessary storage units and other functional equipment including the fireplace/oven are all integrated with the spatial envelope in order to leave the central part of the room empty. So, the center is free from fixed furniture creating the possibility to be used for many different activities with movable interior elements. According to Arat (2011), the integration of necessary storage units and functional equipment with the spatial envelope helps to increase effective use of space. The height of the storage zone is determined in order to ease access for everyday goods

Figure 4. Turkish house interior with the ergonomic spatial concept, the arrangement of ergonomic sitting(sedir), storage zone in accordance with human scale and movements
(Küçükerman, 1989).

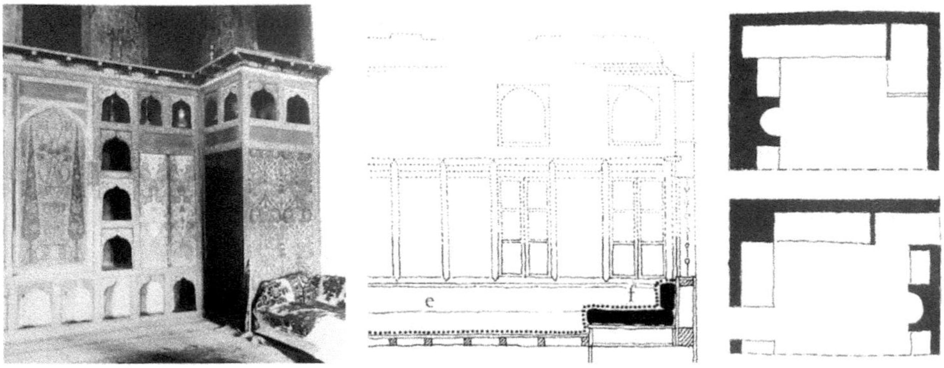

and upper zones difficult to reach are left for decoration. The horizontal distribution of all necessary equipment, the arrangement of continuous horizontal storage zone is in accordance with human scale and movements.

Similar to Turkish house, common traditional Malay House consists of spaces that have their specific uses which in turn influence the interiors and furniture used, making the house a site for many cultural practices and traditions (Hashim, Mohamad Ali & Abu Samah, 2009). Spaces in the house are multifunctional and the use of the space changes during the day and year. There are minimal interior walls and partitions within the house, allowing for flexible use of space and good ventilation and lighting of the interiors. The main spaces and areas in traditional Malay House includes; anjung (varandah), serambi (living area), rumahibu (private area), selang (passageway), and dapur (kitchen) (Yuan, 1987). The Traditional Malays are known for their floor culture, as they sit, eat, work, and pray on the floor (Gibbs, 1987; Zainal Abidin, Khair Ibrahim, Raja Shahminan, 2016). In this sense, for eating, sitting, chatting and many other activities they use traditional mats as body supporting units while they try not to use built-in furniture even for storage. All the equipment and furniture follow the bodily gestures and the ways main activities are fulfilled. This concept increases both the simplicity and flexibility of spatial solutions.

Social Requirements

Social structure includes many different issues all connected to each other such as relationships between people and groups, differences related to lifecycle and family types. These necessitate the consideration of social requirements, which are extremely necessary for subjective well-being. Keyes (1998), defines social well-being as an individual's self report of the quality of his or her relationship with other people, the neighbourhood and the community. The spatial aspects of social well-being are related to qualities that enrich social life. In this sense, interiors should provide the chance for social encounters at desired level for everyone. Moreover they should support and encourage social activities creating a balance between public, semi-public and private places (Saglar Onay, Minucciani, 2018). The quantity and quality of social connections (e.g. talking and listening to family or strangers) correlate with reported well-being as well as physical health. The domestic environment provides the context for this, and the design of our homes and neighbourhoods can facilitate positive social interactions (Green Building Council, 2016).

The social value of vernacular examples includes many tangible and intangible aspects. First of all, the spatial layout of these living environments is strongly connected to the family structure. The family structure plays an important role in the articulation of spaces. For example the African houses of Mousgoum people in Cameroon, Chad and Nigeria called teleukaka are dominated by the family structure and their relations. They are complexes composed of several domestic nuclear structures or hubs for different members of the family and ready to expand with new coming members like daughters in law and children. Interiors of hubs created the necessary privacy for each nuclear family while the henced courtyard was meant to come together for all social and functional needs (Figure5). A similar correspondence can be noted also in the traditional Turkish House. Here rooms take the place of hubs and they support all daily activities of a nuclear family composed of parents and small children. These nuclear families and rooms come together and share a semi-public space called sofa, the gathering place for the whole family composed of sisters, brothers, grandparents and other close relatives who share the same house. Sofa is also used for many different purposes and social activities including engagement, marriage ceremonies, circumcision feasts and big family dinners. So, we can see in these two examples that social relations, values and hierarchy is directly transferred to spatial organization generating a perfect balance between public, semi-public and private space. They are also followers of everyday life giving the possibility to fulfil everyday activities as well as social rituals, which play an important role in strengthening social ties among family members, neighbours and other members of the society. These social relations enhanced by living environments play a crucial role in individual well-being as well as community well-being since in vernacular environments well-being has a more collective character and is enriched by social relations.

Figure 5. The Mousgoum house in Africa reflecting the family structure *(Nelson, 2007).*

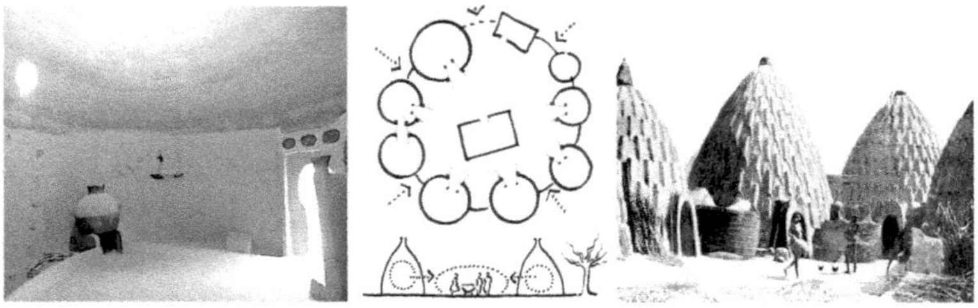

Another social contribution of vernacular architecture is the involvement of the community in the creation of space. This creates an accumulation of knowledge between generations and the community itself owns the knowledge for construction, which evolves by the dynamics of the society creating a strong connection between people and architectural space. Of course this connection is very important and means that the society is capable of building well-being for itself.

Psychological Requirements

Self-determination theorists (Ryan, Deci, 2000; Deci, Ryan, 1985), maintain that well-being hinges on the fulfilment of three basic innate psychological needs which are defined as:

- Competence: Seek to control the outcome and experience mastery
- Relatedness: Desire to interact, be connected to, and experience caring for others
- Autonomy: Desire to be causal agents of one's own life and act in harmony with one's integrated self.

Other than the three basic needs defined by Self-determination theorists, it is necessary to consider two other needs, which are very important in terms of well-being related to the environment. These needs are security and privacy. Security is defined as feelings of basic safety, assurance, and independence from alarm. Privacy is both defined as the state in which an individual's or a group's desired level of social interaction is not exceeded and the right to control (psychologically and physically) others' access to one's personal world, such as by regulating others' input through use of physical or other barriers (e.g., doors, partitions) and by regulating one's own output in communication with others (APA Dictionary of Psychology).

Psychological requirements are the reflections of spatial aspects on human psychology affected also by cultural aspects. In this sense, the requirements may vary greatly from culture to culture. For example the concept of privacy vary greatly among different cultures. A classic Western interpretation on privacy is by Hall (1959), which stated that privacy is about the control of personal space. The Eastern perspectives on privacy differ from the West as the Eastern culture values the privacy of the family as a unit and overriding the importance of individual privacy. In Japanese culture for example, the idea of private spheres that are independent of the group is not acceptable, for the sense of individual privacy exists only weakly (Doi, 1971). The Arab was found to prefer living in large spaces for family congregation and avoid partitions because they do no like to be alone (Hall, 1969; (Hashim, Mohamad Ali & Abu Samah, 2009).

The organisation of space in the Turkish house is dominated by a spatial hierarchy, which helps to acquire the basic three components as well as privacy and security listed above. The concept of the extended family was an important factor in shaping the traditional Turkish house; fathers, mothers, children, sons and daughters-in-law lived under the same roof (Öztank, 2010). Therefore in the Turkish house it becomes necessary to create a room layout that includes all private activities of a nuclear family. Between these rooms there exists a semi-private hall called "sofa" used for getting together for all family members, sometimes neighbours and relatives. In this sense sofa was intended to be a place to socialise during different times of the day. Likely the courtyard was used to come together or perform some activities such as cooking, eating, washing especially for women. It was surrounded by stone walls in order to create the necessary privacy. In this sense, the deliberate interaction between private, semi private and public space, the autonomy of single nuclear rooms, the possibility to control the degree of interaction and the

Figure 6. The traditional Turkish house dominated by a spatial hierarchy, which helps to acquire competence, relatedness and autonomy
(Küçükerman, 1989).

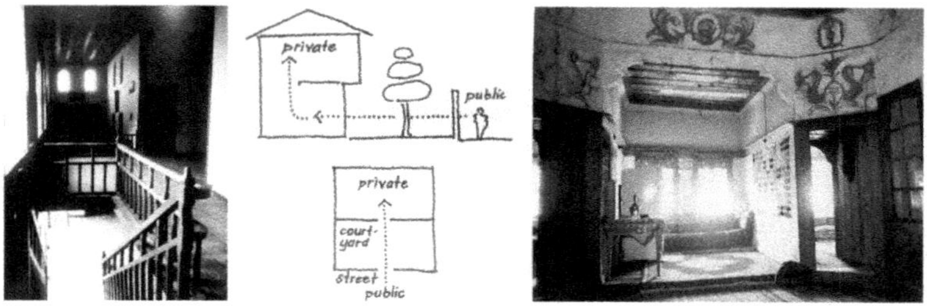

relation of spatial organisation to the family structure were important design criteria in fulfilling psychological needs in the Traditional Turkish House.

In traditional Malay house, serambi (verandah) is used by the men folks for siestas or conversing. For the traditional house in Malaysia, the privacy gradient begins at the most public space which is the anjung (entrance porch), increasing in level to the transition space or semi-public space which is the serambi (reception area) and finally into the most private domain, the rumah ibu (main house). With the introduction of the reception area as a transition zone, social interaction with non-family members were more defined which depends on the type of guests and occasion. The family members have more choice to regulate their privacy and with whom they want to interact within this space. The serambi acts a stop-over zone to mediate between the public space and the more private space. Other semi-private spaces such as the dapur (kitchen) and selang (courtyard) are entered only by family members from the side entrance. During family gathering (khenduri), female relatives and close female friends use the side entrance to go to the kitchen and help with the cooking and food preparations (Hashim, Mohamad Ali & Abu Samah, 2009).

From the examples above, we can see that psychological requirements find their spatial correspondences and they are respected by all members of the society in their interactions. Spatial organization and architectural features like room order, passage types, character of openings are all concrete reflections of intangible values related to psychological needs. An absence of these concepts in spatial terms may cause severe psychological problems and affect well-being of inhabitants.

Sensory Requirements

Sensory requirements are extremely effective on well-being and they are concerned with both quantitative factors like heating, illumination, acoustics, sound control etc. and their reflections on human senses. Pallasma argues that (2005) every touching experience of architecture is multi-sensory; qualities of space, matter and scale are measured equally by the eye, ear, nose, skin, tongue, skeleton and muscle. Relatedly Zumthor (2006) states that we perceive atmosphere through our emotional sensibility, which is a form of perception that works incredibly quickly. We are capable of immediate appreciation, of a spontaneous emotional response. So sensory requirements affect our feelings both in physical and cognitive levels.

A key element of achieving mental well-being is the element of control. When people feel in control of their surroundings and environment, they feel empowerment and stability (Shapiro, 1999), (Taylor et al, 1999). In physical means, sensory requirements needs to be controlled by the solutions created both in architectural and interior levels. The architectural language in vernacular homes are usually responsive to all aspects related to climate and topography. They also include secondary control mechanism that can be adjusted by users themselves. Sometimes they are based on solutions that make it possible to use space in different ways according to changing seasons. The Turkish house responds to climate changes with its overall design and connection to contextual aspects. By employing the substantial knowledge accumulation, both the spatial organization and orientation of rooms and the use of local materials and construction techniques were effective in creating simple but very clever solutions. Küçükerman (1989) points out that houses in Turkish settlements were provided with spaces both 'open-cool' and 'closed-well protected' in accordance with different seasonal requirements. Küçükerman (1989) describes these spaces: the 'open-cool' summer room is a space that remains relatively cool by being situated in a direction that is open to air flows, and with thinner walls and floors, and wide and large windows. The 'closed-well protected' room, on the other hand, is protected from flowing air, is situated in middle floors, and built with thick and non-permeable materials and with small windows.

Similar to the traditional Turkish house, the Iranian vernacular architecture responding to hot and dry climate has a flexible understanding of space use according to seasons. In these houses different rooms present different thermal environments to the occupants who can select the most comfortable condition by moving from one room to another. North facing summer rooms are least exposed to the sun compared to the other rooms. In fact, exposure occurs only in the early and late hours of summer days when the angle of altitude is low and the angle is such that the sun's rays are almost tangential to the surface of the wall (Foruzanmehr, 2006; Foruzanmehr, 2016). They also have high ceilings for the hot air to rise up and leave the living level cooler. Moreover their thick walls and roofs retard and reduce the transmission of heat. Heavy structures store and retain heat during the day and release it in the evening and at night (Bonine 2000; Foruzanmehr, 2006; Foruzanmehr, 2016). On the other hand, south-facing winter rooms are planned so that the low rays of the winter sun can shine through their windows. The sunshine passes through the glass of timber doors, and its heat is stored in mud brick walls, which are about 70 cm thick. The heat is released gradually within 7–9 hours to keep the family warm when the sun moves on (Beazley and Harverson 1982; Foruzanmehr, 2016). Winter rooms have lower ceilings compared to summer rooms to help warm the rooms more easily and quickly. In very cold winters, the sun's energy is not sufficient, and extra energy needs to be used to warm up the winter rooms. In order to reduce the energy consumption on winter nights, a smaller place for sitting and sleeping is heated instead of warming the whole winter room (Foruzanmehr, 2016).

The Malay house can be another good example for hot and humid climate with increased need for ventilation. A traditional Malay House is a timber house standing on stilts letting the wind pass under the house. Windows are plentiful, lining the walls and providing good ventilation and views for the house. There are minimal partitions or interior walls in the house, allowing for the flexible use of space. According to Yuan (1987), a house form in Malaysia should allow adequate ventilation for cooling and reduction of humidity; use building materials with low thermal capacity so that little heat is transmitted into the house; control direct solar radiation; control glare from the open skies and surroundings; protect against rain; and ensure adequate natural vegetation in the surroundings to provide a cooler micro climate. Solar radiation is controlled with the large thatched roof while glare is controlled by excluding open skies and bright areas from the visual field creating under lighted interiors which gives a psychological effect of coolness as strong light is often mentally associated with heat (Yuan, 1987).

Light is essential for well-being. There is a general perception that daylight should be the primary light source for health and well-being (Veitch, Hine & Gifford, 1993; Christoffersen, 2011). On an emotional level, people enjoy and feel a sense of well-being in daylight and sunlight. Natural light makes

Figure 7. The Malay house efficiently controlling the severe effects of the hot and humid climate
(Yuan,1987).

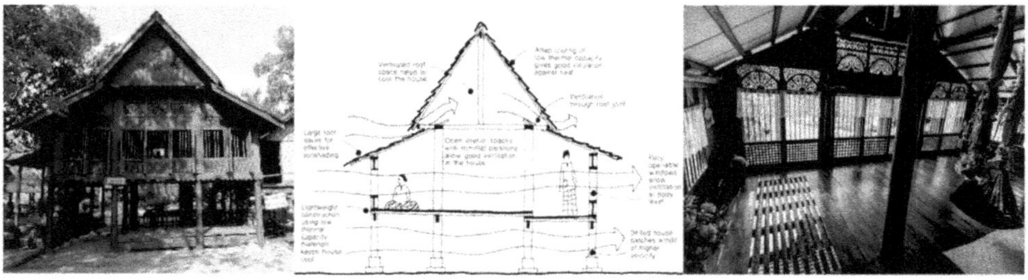

us aware of the passage of time: the specific mechanisms of this interaction are not entirely understood, but our bodies are naturally in tune with external light levels and characteristics (Green Building Council, 2016). At this point vernacular architecture is traditionally based on the natural fulfilment of the need for light, as systems of artificial lighting were not very common by the time they were created.

In vernacular examples the proper use of natural light brings about incredible health benefits effecting all dimensions of well-being including contextual, functional, psychological, aesthetic and social aspects. The Iranian traditional house handles the issue of natural light in a multi-beneficial way. First of all, the proper orientation of the whole building is coordinated to use sunlight more effectively during winter and avoid the extra sunlight in summer (Makani, Khorram & Ahmadipur, 2012). The dimensions of the openings are adjusted according to the direction, size and purpose of the space it will supply light, heat and ventilation. The existence of open spaces like cortiles, closed spaces and semi-sheltered spaces around the cortiles create a variety spaces with different degrees of exposure to the sun and provide the conditions for doing the daily routines. This also enhances physical and psychological health. Views provided by the openings also contribute to social and psychological well-being. And moreover the colourful pieces of glass create an atmosphere that changes with daylight and contributes to spatial aesthetics. Revealing the aesthetical aspect of light, these features create a spiritual atmosphere by producing attractive light effects inside buildings (Pirnia, 2006). According to Afshar Naderi (2003), light beams penetrating the interior space through these windows often create remarkable patterns on the wall or the floor, which looks like an artistic painting or a precious carpet.

The underlined features that develop in relation to sensory data also play an important role in creating unique spatial identity. The context related data gives the possibility to create regional architectural identity and differentiates from other examples in different contexts. As we can see in the vernacular examples discussed above, the architectural and spatial solutions created by responding to the character of a specific environment have a lot of benefits regarding sensory input.

Aesthetic Requirements

Wolfgang Welsch (1998) argues that 'the aesthetic' is a polysemy in that there is a wide variety of usages of the term circulating which, although inter-related, do give one quite distinct perspectives on the topic. Some of these are as follows:

- The ability to make a harmonious appealing whole from disparate elements;
- The ability to perceive contrasts between contiguous elements,
- The appreciation of the sensuous – that which appeals to all the senses;
- The appreciation of that which requires the higher cultivated senses;
- That which requires perceptiveness rather than sensateness;
- That which requires time to appreciate and is beyond the immediacy of the moment;
- That which concerns itself with phenomenological appearance and not substance, and
- The ability to draw all the above elements into one piece of artistic creation (Welsh, 1997; Dale K. & Burrell G. (2003).

In defining the relation between architecture and aesthetics, two different conceptions are considered. In the pragmatic view, exact parameters are accepted as self-sufficient (symmetry, rhythm, proportion...) and they define architecture as aesthetic object, while in the idealistic view those exact parameters represent just tools of non aesthetic values through which they achieve necessary sensual appearance of the ideas (Jovanovic-Popovic, Sunjkic & Tomovska, 2012). Idealistic concept is an attempt to establish essential values that belong to field of aesthetics, and that are observed through philosophical, cultural, religious, sociological aspect (Mako, 2005). Traditional aesthetics is based on pragmatic aesthetic concept, while contemporary aesthetic interpretations are based on idealistic concept (Jovanovic-Popovic, Sunjkic &Tomovska, 2012).

In spite of the expansive scope of environmental aesthetics, a basic assumption of the field is that every environment, natural, rural, or urban, large or small, ordinary or extraordinary, offers much to see, to hear, to feel, much to aesthetically appreciate (Carlson, 2000). It also focuses on smaller and more intimate environments, such as our backyard, our office, and our living room (Carlson, 2000). So, in every spatial experience, we tend to make aesthetic evaluations. When these evaluations are positive, the quality of spatial experience is enriched and thus the pleasing effect of space on our senses is increased. In other words, in our experience of space, there is always the aesthetic dimension and this has a great effect on our well-being related to space.

According to Jovanovic-Popovic, Sunjkic and Tomovska (2012), vernacular house can be characterized as material form based on harmonic relations between a particular person and a particular place. Harmony between house form and natural and generated environment is not simply the results of physical forces or any single casual factors, but in the consequence of a whole range of socio-cultural factors seen in their broadest term. Listening to the nature, and profound considering of the concrete conditions of the environment are very significant for creating the quality architecture (Jovanovic-Popovic, Sunjkic & Tomovska, 2012). Departing from this view we can say that the appropriate relation with nature and natural sources creates an aesthetic quality similar to the one we see in the natural environment. In this sense living environments in vernacular settings become an extension of nature.

Aesthetics in interior space also needs to be accepted as a very important quality that we can not do without. Aesthetic aspects affect the way people feel depending on spatial aspects, furniture and objects. On the other hand, the understanding of aesthetics is strongly based on cultural aspects. In vernacular examples aesthetics is usually connected to other functional, ergonomic aspects as a natural outcome of clever solutions, the nature and capacity of materials and the way spatial elements and objects are created. For example the aesthetic appearance of Japanese room is characterised by the strong relation between the room and the garden, which becomes an extension of the interior. The façade has many large door and window apertures allowing the rooms to open directly to the outside and creating a strong relation with nature (Yoshida, 1952). The Japanese veranda is an architectural aspect that characterizes the relation between inside and outside. The modular understanding of space with tatami, the partitions with ornamental transoms and decorative shelves are all spatial elements that come together in harmony creating a strong aesthetic identity.

Figure 8. The aesthetic identity in the Japanese Vernacular House (Minka), the strong relation with the garden, modular spatial elements and elaborate ornaments (Kawashima, 1986).

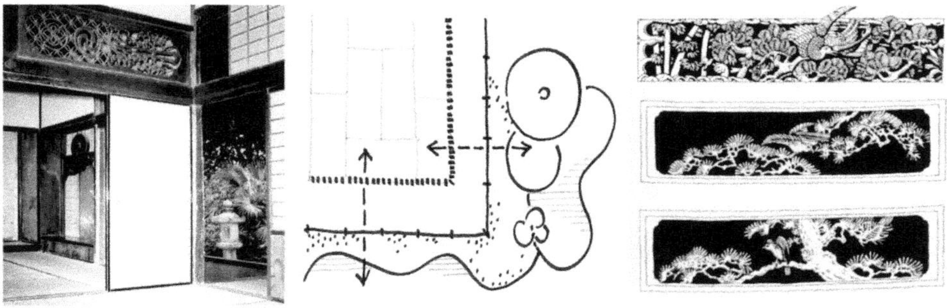

Cultural Differences

The human factor is one of the most important determinants of vernacular architecture as they are the mirrors of ways of thinking of their inhabitants who have also built them. According to Christian Norberg Schulz (1985), the purpose of architecture is to accommodate a human being to dwell in his world so that he may stay poised in place and time. To be able to establish a meaningful relationship with a given environment, it is necessary that architecture goes beyond neither stylistic elements nor abstraction concept in its effort to manifest the dwelling concept. In this sense vernacular houses are expressions of cultural values generated through the relation between the natural environment and people. The arrangement and design of Batak Toba houses in Sumatra reflect the cultural coherence and deep-rooted tradition of the people. According to Hanan (2011), a Batak Toba house embodies existential meanings of its inhabitants and the staying between the earth and the sky. The earth is alive with active forces and people display respect and awe to those. The world is created by God in three manifestations: one is governing the upper world (God's place), one is governing the middle world (place for the human being) and one in the underworld (the place of demon and spirit). The house respectively is divided in three sections: the lower part for the storage of wood and animals which symbolize the foot, the middle part as the house of the human being which symbolize the body, the attic constituting the biggest part of the house, as the upper world for the God or ancestors which symbolize the head (Hanan, 2011).

Figure 9. Batak Toba House shaped by the cultural values, using human analogies as a metaphor for comparable spatial uses, structural, decorative and symbolic forms (Hanan, 2015).

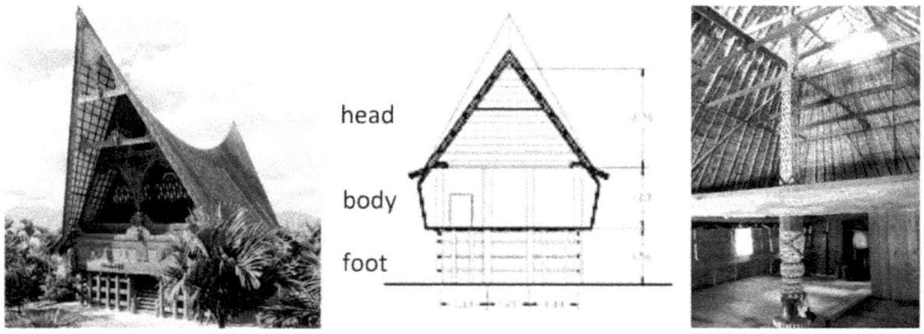

Cultural differences in fact affect all the spatial requirements discussed above. What can be concurred is that the built environment can best be seen as a setting for behaviour and it affects behaviour without determining it. People do behave and act differently in different behaviour settings because they want to match or make congruent their behaviours appropriate to the setting as defined by culture and social norms (Hashim, Mohamad Ali & Abu Samah, 2009).

FUTURE RESEARCH DIRECTIONS

Through the examination of vernacular homes, we have seen that the requirements defined in the well-being framework all find their spatial correspondences. In other words we can say that vernacular buildings develop as a response to the real needs of a specific society also affected by the relationship between the society and context. Today we are very far away from this point, as we designers and architects tend to create architectural solutions without a proper examination of ways of living. It is also a reality that the difference between cultures and lifestyles is not as significant as before but they are adapted to the advancements in technology and science. In this sense the mission of design and architecture is to understand this adaptation process so that the contemporary ways of living and thinking can find their correspondences in new living environments. So, the approach of vernacular examples becomes important at this point. Future research needs to discuss how new architectural solutions can respond to well-being needs in a natural way as we can see in vernacular examples and how we can develop an updated and contemporary understanding of vernacular architecture. This is also because our present ways of living is excessively impactful on the environment and living environments are no more extensions of nature. As a result, the built environment does not respect nature and consequently nature does not respect the built environment.

CONCLUSION

Conservation of vernacular architecture is an important issue to inspire future generations teaching us how to be integrated with the environment by taking the advantage of local sources. This understanding has to be a part of architecture and design education in order to discuss their exceptional design approaches and to adapt them to new living environments. Therefore conservation of vernacular architecture as a lesson to preserve a sort of "biodiversity" in

Figure 10. The concept of conservation of vernacular architecture for the interior architecture body of knowledge

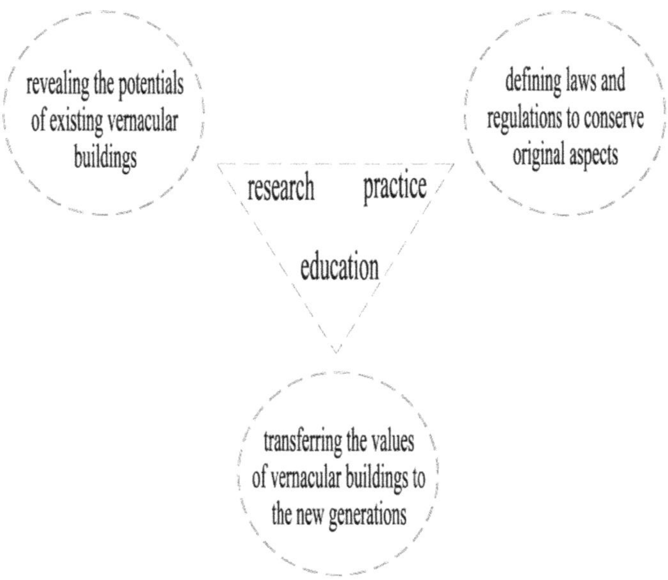

lifestyles, can contribute greatly to design education, research and practice. Vernacular houses represent an irreplaceable value for the interior architecture body of knowledge as they are greatly connected to cultural aspects distinctive from society to society far away from the impacts of "Global" or "International Style" that create similar solutions in different contexts.

Another important issue to appreciate vernacular architecture is to conserve interiors as mirrors of everyday life in times when they are created. This necessitates the re-evaluation of laws and regulations regarding the authenticity of interiors.

Thus conserving vernacular architecture and interiors need to be evaluated in three basic steps including research, revealing the potentials of existing vernacular buildings; practice, defining laws and regulations in order to conserve buildings and interiors with their original aspects; education, transferring the values of vernacular understanding to the new generations. New design approaches need to consider both the potentials of contextual relations and human ways of living and requirements, which will lead to a significant contribution to human well-being in everyday living environments.

REFERENCES

Afshar Naderi, K. (2003). *Iranian Architecture*. Tehran: Aghaah.

APA dictionary of psychology. (n.d.). American Psychological Association.

Arat, Y. (2011). *Analysis of traditional Turkish house indoor components based on antropometric data: Konya houses* (PhD Thesis). Seljuk University Institute of Science and Technology.

Beazley, E., & Harverson, M. (1982). *Living with the desert: Working buildings of the Iranian plateau*. Aris & Phillips.

Bonine, M. (2000). Sustainable desert housing: From the dwelling to the desert community. In *Proceedings of Sustainable development of desert communities. Regional Symposium*. Yazd: UN Development Programme.

Carlson, A. (2000). *Aesthetics and the environment: The appreciation of nature, art and architecture*. Routledge.

Christoffersen, J. (2011). The importance of light to health and well-being. In *Proceedings of 4th VELUX Daylight Symposium "Daylight in a Human Perspective"*, Lausanne, Switzerland: Rolex Learning Center École Polytechnique Fédérale de Lausanne (EPFL).

Correia, M., Carlos, C., & Rocha, S. (2014). *Vernacular heritage and earthen architecture: Contributions for sustainable development*. CRC Press, Taylor and Francis.

Dale, K., & Burrell, G. (2003). An-aesthetics and architecture. In A. Carr & P. Hancock (Eds.), *Art and Aesthetics at Work* (pp. 155–173). Palgrave Macmillan. doi:10.1057/9780230554641_11

Deci, E. L., & Ryan, R. M. (1985). *Intrinsic motivation and self-determination in human behavior*. Plenum. doi:10.1007/978-1-4899-2271-7

Doi, T. (1971). *The anatomy of dependency*. Kobundo.

Foruzanmehr, A. (2006). *Energy efficiency and sustainability in vernacular architecture of Iran: Traditional design, construction methods and passive strategies* (MSc diss.). Oxford Brookes University.

Foruzanmehr, A. (2016). Thermal comfort and practicality: Separate winter and summer rooms in Iranian traditional houses. *Architectural Science Review*, *59*(1), 1–11. doi:10.1080/00038628.2014.939132

Gibbs, P. (1987). *Building a Malay house*. Oxford University Press.

Green Building Council. (2016). *Health and well-being in homes*. Green Building Council.

Hall, E. T. (1959). *The hidden dimension: Man's use of space in public and private*. The Bodley Head Ltd.

Hall, E. T. (1969). *The silent language*. Doubleday and Company, Inc.

Hanan, H. (2011). A house is a figure between the Earth and the Sky Case Study: Batak Toba House in Samosir Island. *RUAS*, *9*(2), 51–60. doi:10.21776/ub.ruas.2011.009.02.7

Hashim, A. H., Mohamad Ali, H., & Abu Samah, A. (2009). Urban Malays' user behaviour and perspective on privacy and spatial organization of housing. *Archnet-IJAR*, *3*(1), 197–208.

Hidayahtuljamilah Ramli, N. (2012). Re-adaptation of Malay house thermal comfort design elements into modern building elements: Case study of Selangor traditional Malay house & low energy building in Malaysia. *Iranica Journal of Energy & Environment 3*, 19-23.

ICOMOS. (1999). *The Charter on the Built Vernacular Architecture*. Retrieved from www.icomos.org

Jovanovic-Popovic, M., Sunjkic, V., & Tomovska, R. (2012). Aesthetics of vernacular architecture comparative analyses of context aesthetics in Balkan Region. In *PLEA 2012 28th Conference, Opportunities, Limits & Needs Towards an environmentally responsible architecture*. Lima: PLEA.

Kawashima, C. (1986). *Minka: Traditional houses of rural Japan*. Kodansha International Ltd.

Keyes, C. L. M. (1998). Social well-being. *Social Psychology Quarterly, 61*(2), 121–140. doi:10.2307/2787065

Küçükerman, Ö. (1989). *Kendi Mekanının Arayışı İçinde Türk Evi [Turkish House in Search of Spatial Identitiy]*. Türkiye Turing ve Otomobil Kurumu.

Makani, V., Khorram, A., & Ahmadipur, Z. (2012). Secrets of light in traditional houses of Iran. *International Journal of Architecture and Urban Development, 2*(3), 45–50.

Mako,V. (2005). *Estetika-Arhitektura: Sedam tematskih rasprava*. Belgrad: Orion Art.

Mako, V. (2012). Aesthetics in architecture: Contemporary research issues. *SAJ, 4*, 134–141.

Nelson, S. (2007). *From Cameroon to Paris. Mousgoum architecture in & out of Africa*. The University of Chicago Press.

Norberg-Schulz, C. (1985). *The concept of dwelling. On the way to figurative architecture*. Rizzoli International Publications.

Öztank, N. (2010). An investigation of traditional Turkish wooden houses. *AABE, 9*(2).

Pallasma, J. (2005). *The eyes of the skin*. John Wiley &Sons.

Pheasant, S. (1991). *Ergonomics, work and health*. Palgrave Macmillan. doi:10.1007/978-1-349-21671-0

Pirnia, M. K. (2006). *An introduction to Islamic architecture of Iran*. Sorooshe Denesh.

Rael, R. (2009). *Earth architecture*. Princeton Architectural Press.

Ryan, R. M., & Deci, E. L. (2000). Self-determination theory and the facilitation of intrinsic motivation, social development, and well-being. *The American Psychologist*, *55*(1), 68–78. doi:10.1037/0003-066X.55.1.68 PMID:11392867

Saglar Onay, N., & Minucciani, V. (2018). *Well-being framework as a contributor to sustainability. WIT Transactions on Ecology and the Environment, 217.*

Salvendy, G. (2012). *Handbook of human factors and ergonomics*. John Wiley & Sons. doi:10.1002/9781118131350

Santomauro, A. (2014). Come ristrutturare una vecchia cascina e darle nuova vita. *Design Mag*. Retrieved from www.designmag.it

Shapiro. (1998). *Control therapy*. Wiley.

Taylor. (1998). Illusions and well being: A social psychological perspective on mental health. *Psychological Bulletin*, *103*, 293–210.

UNESCO. (n.d.). *City of Safranbolu*. World Heritage List. Retrieved from https://whc.unesco.org/en/list/614

Veitch, J. A., Hine, D. W., & Gifford, R. (1993). End-users' knowledge, beliefs, and preferences for lighting. *Journal of Interior Design*, *19*(2), 15–26. doi:10.1111/j.1939-1668.1993.tb00159.x

Welsch, W. (1998). *Undoing aesthetics*. Sage Publications.

Yoshida, T. (1952). *The Japanese house and garden*. Pall Mall Press.

Yuan, L. J. (1987). *The Malay house*. Institute Masyarakat.

Zainal Abidin, N., Khair Ibrahim, F., & Raja Shahminan, R. N. (2016). Re-imagining sustainable furniture by using ergonomic applications of the traditional Malays: A review. In *Proceedings of ICSEMSS 2016*. Johor Bahru: Universiti Teknologi Malaysia.

Zumthor, P. (2006). *Atmospheres*. Birkhauser.

Chapter 10
Cultural Heritage and Adaptive Reuse as a Contributor to Well-Being

ABSTRACT

This chapter aims to discuss the dialectic relationship between interior environments of heritage buildings and users and the importance of preserving interior elements as communicators of cultural significance. Today the conservation practice and reuse proposals focus on preserving the architectural envelope of buildings rather than interior aspects. Conversely, interiors need specific safeguard and care, not only due to their cultural significance but also because they are the part most closely connected to the real life of users. In the perspective of cultural conservation, existing buildings represent the continuity of cultural values from the past to the future, and at this point, interior architecture is crucial as it provides an authentic interaction between users and spaces conveying all these values. Therefore, the conservation of interior envelope and interior elements rises as a very important issue to be discussed, affecting the occupant's well-being in a very subjective and sometimes unconscious way (considering the cultural belonging).

DOI: 10.4018/978-1-7998-4231-6.ch010

INTRODUCTION

"Reuse" is defined as "making suitable for new requirements through modifications" (Burden, 2004). According to the Venice Charter of 1964 the conservation of buildings is always facilitated by making use of them for some socially useful purpose but only modifications demanded by a change of function is permitted (ICOMOS, 1964). The Venice Charter proposes 4 main requirements to be taken to account for conservation and restoration of historic buildings, which can be summarized as authenticity, reversibility, compatibility and retreatability.

The authenticity of a structure can be defined as different layers of significance. The present situation of an antique structure even destroyed or damaged represents a "state of understanding" (Sickels-Taves, Lauren, 1999). Due to the requirement of the preservation of as much of the original material as possible, any intervention should be kept minimum. This means that nothing more should be done than is strictly necessary to guarantee the proper use, conservation and prolongation of the life of the original fabric (Sickels-Taves, Lauren, 1999). Conservation theory has also placed value on "reversibility" which is defined as the ability to undo repairs that later generations regard as substandard or incorrect (Winter, 2007). For interiors reversible interventions are even more important as interiors are changed more often according to changing human needs. Differentiation of the new and the old helps to maintain reversibility since additions are often structured as independent bodies and they are built of different materials and techniques with a new design language. Standards for Rehabilitation published by "U.S. Secretary of the Interior" in 1978 strongly propose that the new work should be differentiated from the old and it should be compatible with the historic materials, features, size, scale and proportion. SBAP principles predominantly used within Great Britain encourage expression of modern needs in modern language (Sickels-Taves, Lauren, 1999). The last requirements, compatibility aims to ensure that introduced materials will not have negative consequences and retreatability focuses on the type and extent of conservation treatment in order to make sure that the present interventions would not preclude or impede further treatment of the future.

The historic environment is subject to change including buildings and all their interiors and adaptation is an inevitable outcome of this process. According to the Burra Charter (ICOMOS, 2013), adaptation means changing a place to suit the existing use or a proposed use. Use means the functions of a place, including the activities and traditional and customary practices that may occur at the place or are dependent on the place.

Spatial potentials are mostly defined by the innate and unchanging properties of the building and they are very important in order to understand the cultural significance of the place. The interior environments of heritage buildings can have different meanings and cultural significances according to the building typology and relationship between architectural envelope and interiors. In this sense the appropriate way of proposing a new use might differ a lot from building to building. But the most important concern is to consider the potentials of buildings and their surroundings as well as the expectations and needs of users. A recent architectural competition in Turkey was titled "what would the Izmir Electric Factory would like to become?" aiming to increase awareness about the values of heritage buildings in collective memory and public life. In this sense, interiors of buildings with specific cultural value have a great potential as communicators of cultural significance. Buildings representing different historic and cultural layers can contribute to the continuation of public memory only if their interiors can communicate the embedded meanings about their existence. (Saglar Onay, Minucciani, 2019).

Cities and all their components including buildings and their interiors are living structures. If they are abandoned without a use, they become frozen in time. Reuse is the way to make them breathe and make them apart of our lives (Saglar Onay, 2014).

ADAPTIVE REUSE AND COMMUNITY WELL-BEING

The integration of historic conservation with environmental concerns has become an innate feature of an agenda to support sustainability (Stubbs, 2004; Bullen and Love, 2010). As part of a wider revitalisation strategy to promote sustainability within the built environment, many buildings of cultural and historical significance are being adapted and reused rather than being subjected to demolition (Ball, 1999; DEH, 2004; Wilkinson and Reed, 2008; Wilkinson, Reed & Kimberley 2009; Bullen & Love, 2009; Bullen & Love, 2011).

According to McCallum (2007), reusing buildings is a simple way of achieving sustainability and it helps to reinforce a strong sense of place. Especially for people who live in the same environment for a long time, sense of place and place attachment has a great impact on well-being. Place attachment is defined as a feeling of comfort, which the individual is benefited from in his or her living place (Brown, Perkins, & Brown, 2003) and is considered as a strong predictor for social well-being especially in older adults' population (Rollero & De Piccoli, 2010).

Place attachment in personal context consists of components that explain the cognitive relationship between the person and his or her physical environment. Proshansky (1978) stated that place identity refers to a person's identity based on his or her values, goals, and skills in connection with the place (Kyle, Graefe, Manning, & Bacon, 2004).

Reuse of heritage buildings is mostly discussed in terms of its benefits on the environment and its contribution to the existing fabric of cities. On the other hand the benefits of reuse on the society is even more significant. According to McCallum (2007), when the historic environment is evaluated appropriately, it contributes to the quality of life and enriches people's understanding of the diversity and changing nature of their community and it becomes a powerful focus for community action. Also according to the Department of Environment and Heritage (DEH, 2004) heritage buildings provide a valuable glimpse of the past and lend character to communities and therefore should be conserved for future generations. The existence of living historic buildings and sites create a platform to come together and share common values of the society. This can be regarded as a booster of social well-being for all levels of the society starting from the people who can share memories about the past of these buildings and also including the younger generations who can explore concrete values about the society they live in. The existence and sharing of these important cultural values strengthen social ties and belonging. In this sense, the appropriate evaluation of historic buildings through adaptive reuse needs to be regarded as a very important contribution to well-being. Looking from this perspective, well-being is not a concept that can be evaluated only for the individual but for also for the society. Individual well-being is not a concept independent from the community. The understanding of community well-being is concretely supported by the environment that respects common values of the society. In this sense adaptive reuse of buildings for social purposes is strongly linked with community well-being which in turn directly affects individual well-being.

The concept of community well-being is linked to also relatedness, which can be regarded as one of the basic innate psychological needs. Self-determination theorists (Deci, Ryan, 1985, Ryan, Deci, 2000), maintain that well-being hinges on the fulfilment of three basic innate which are defined as: competence: Seek to control the outcome and experience mastery;

relatedness: desire to interact, be connected to, and experience caring for others and autonomy: desire to be causal agents of one's own life and act in harmony with one's integrated self. Relatedness is a feeling of connection with other people, often accompanied by affection, trust, and a sense of personal security (APA Dictionary of Psychology). So, relatedness can be also regarded as a social dimension of psychological needs. People need to care for other people and they need to be cared about by others. From a cultural, more precisely a societal or country point of view, the conceptions of individualism and collectivism have become pervasively prominent (Triandis 1995). In this conception, individualism relates to autonomy, whereas collectivism relates to relatedness (Kellert, 2016). At this point, it can be assumed that today's societies need to enhance the feeling of relatedness, which has been dominated by the effects of individualism. Oliver (2020), discusses that the age of the individual must end and our world depends on it. The more we see ourselves as discrete entities, the more likely we are to feel isolated and lonely and to show "selfish" behaviours. As a consequence, rates of anxiety and depression are rising across the world, while the climate and biodiversity crises deepen ever further.

DIFFERENT CONCEPTIONS OF REUSE FOR INTERIORS

Depending on the architectural layout and character of the envelope, interiors can have different meanings and priorities. The type and extent of reuse proposals can vary greatly according to these different conceptions.

- Interiors dominating architectural and spatial Identity
- Interiors as followers of architectural envelope
- Interiors as spatial components
- Interiors perceptually detached from the architectural envelope
- Interiors reflecting particular aspects and elaborate detail
- Urban interiors (exterior interiors)

Interiors Dominating Architectural and Spatial Identity

Depending on the character of architectural layout and envelope, architectural identity can be highly dominated by the interior environment. In this case, the interior dominates spatial perception and experience. The Grand Bazaar can be a very significant example to this type of spatial development (Figure1). The bazaar cannot be perceived as an entire building from the exterior while the interiors underline all the specific aspects of spatial organization and identity. The Old Bedesten of Grand Bazaar can be regarded as the most precious part of the bazaar as it was also used for storing the most valuable merchandise in the bazaar. It is one of the oldest examples of the Bedesten typology, which is a commercial building usually composed of single commercial units surrounding a central space. The central courtyard is covered by 15 domes resting on pointed arches supported by 8 tetragonal bases. Nevertheless in time this courtyard was divided into smaller parts and sold to individual merchants. Today the courtyard is dominated by one-storey commercial units that are

Figure 1. The vaults as architectural elements to be underlined in Old Bedesten situated in Grand Bazaar
(Saglam, 2015).

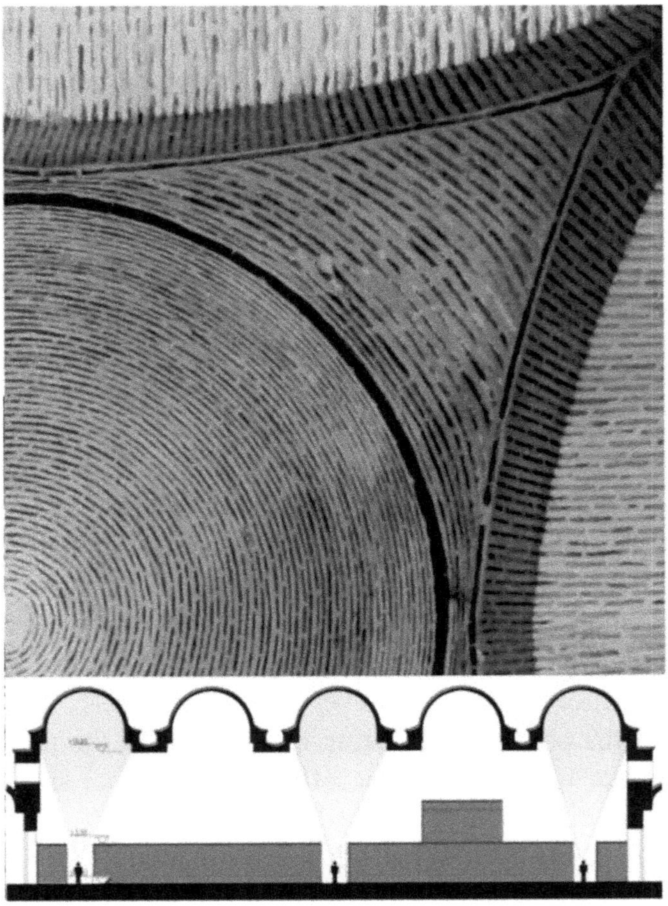

used to exhibit antique Ottoman goods. Even if the units can physically be interpreted in different ways, they cannot be removed. Therefore these different physical layers that have become part of building's identity, restrict greatly the type of possible interventions. While evaluating different spatial layers in the Old Bedesten, we can observe a chaotic situation. Today when we go there we cannot see the vaults, which characterise the interior envelope of the building because they are partially hidden by the dominance of the retail units. Figure 1 shows the physical layers that block perception of the original spatial envelope. The section at the bottom part shows part of the void that cannot be perceived by the observer. While dealing with historic buildings with important cultural background and high architectural quality, it is not enough to preserve the physical aspects but it is also essential to make them a part of spatial experience. In the example of Bedesten the interior envelope is the most important spatial aspect to be underlined. The original brick vaults, arches and walls are elements that define spatial identity so indisputably they are to be intensely emphasized. (Saglar Onay, Minucciani, 2019).

Another example for the dominance of Interior Space is the "Palazzo dell'Esposizione Internazionale del Lavoro", or simply "Palazzo del Lavoro" designed by the engineer Pier Luigi Nervi (in collaboration with architect Gio Ponti and Gino Covre) as part of the large complex of "Italia 61" built for the celebrations of the Unification of Italy centenary: it is a compact volume that overlooks an urban artery of great flow, at the southern gates of the city (Figure 2). In this case the interior atmosphere created by the structure of the building is strongly legible from the interior and not from the exterior (except at night, when – if illuminated - the interior appears). The pillars and their expressive connection to the roof structure create a unique pattern and spatial identity to be underlined. The outer shell is technologically interesting but - especially after so many years of neglect - not at all attractive, it hides the spectacular interior: huge "tree pillars" of 40 m side separated by continuous strips of skylights mark an empty space totally disproportionate to the human dimension and therefore all the more surprising. Three mezzanine floors have been created along the perimeter, overlooking the enormous central void. It was a remarkable exhibition structure in terms of size and technological innovation, but after some exhibition events the problems related to its size prevailed over the possibility of using it continuously. Its use was varied, as it hosted trade fairs, exhibitions, offices, educational spaces, sporadic commercial activities, to be abandoned in time (Chiorino, 2015). In 2015 a fire made the situation even worse, but also drew attention to the issue of its future: over the years, a number of projects and proposals have taken turns,

Figure 2. The interior of Palazzo del Lavoro, Torino
(https://www.torinostoria.com)

also trying to involve the inhabitants. The difficulty of preserving and at the same time restoring life to this volume is the reason why it is still unused today. In 2008, a Ltd. company bought the Palazzo del Lavoro to turn it into a shopping mall: the project was discussed, rejected, re-proposed several times and now, after the withdrawal of Saudi Fawaz Alhokair who were supposed to invest in this operation, works cannot yet start. Regarding future interventions, to avoid misrepresenting the sense and perception of interior space is the main concern, to ensure a true preservation of this extraordinary architectural heritage. (Saglar Onay, Minucciani, 2019).

As we can see from the two examples from two different cultural settings, the interiors of heritage buildings with significant value are subject to inappropriate interventions and evaluations. Sometimes they are not integrated to social life and sometimes they are not evaluated correctly in terms of spatial experience. At this point it is important to underline that interiors have a great potential to contribute to social life and community well-being. The Grand Bazaar is an interior environment that continuously communicates cultural values of the past both to the community rooting from the city and to other communities of the world. In this sense, interiors with specific spatial identity need to be evaluated for communal activity together with all their spatial layers of cultural significance.

Interiors as Followers of Architectural Envelope

Some interiors have specific importance and value because of their relatedness to the architectural envelope. In this case, interiors are mostly defined by the structural system and architectural shell of the building. On a different scale, the famous Gaudì's Milà house in Barcelona (1906-12) can be a good example for this kind of interiors. The extraordinary masterpiece of the Catalan architect will not be discussed here, but rather an intervention carried out in its attic: under the roof, spaces were originally used as storage and laundry rooms, but in the 1950s the Catalan architect J. Barba Corsini was commissioned to create a series of apartments on this floor, which he could design by masterfully exploiting the opportunities offered by the structure (based on parabolic arches of partition wall) and its sinuous layout (Ruiz Millet, 1995; Carabi-Bescós, 2018, Trucato) Playing with the increasing heights, Barba Corsini designed 13 duplex apartments. To access the apartments, a new elevator was installed directly from the ground floor, while in the exterior he opened small windows same as pre-existing ones, as well as new building ventilation chimneys, indistinguishable from the original ones. This project is a true "interior architecture", exploiting existing spaces and shapes. We have several black and white photos and drawings, which testify very interesting solutions, typical of the modern period (Figure 3), (Saglar Onay, Minucciani, 2019).

Unfortunately, In the name of a "purist" (but actually obtuse) conservation, when Pedrera was acquired by the Catalunya-La Pedrera Foundation to house a cultural center, these apartments, as unrelated to the original project, were dismantled bringing the attic back to its previous state. Two considerations must be taken into account: the Pedrera conservation involved the dissolution of another work of architecture, specifically and exclusively internal, so there was a conflict between two conservation instances that probably could be avoided; secondly, the conservation of the attic "in the original state" is actually an illusion: in fact, the true original state of the attic was a simply service space, taken from the view of visitors, while now it has acquired a more noble role that did not belong to him, being an integral part of the visit: it now houses an exhibition gallery, called Space Gaudì, in which the work of the architect is presented. (Saglar Onay, Minucciani, 2019).

Figure 3. Interior of La Pedrera apartments, by the Catalan Architect J. Barba Corsini (Ruiz Millet, 1995).

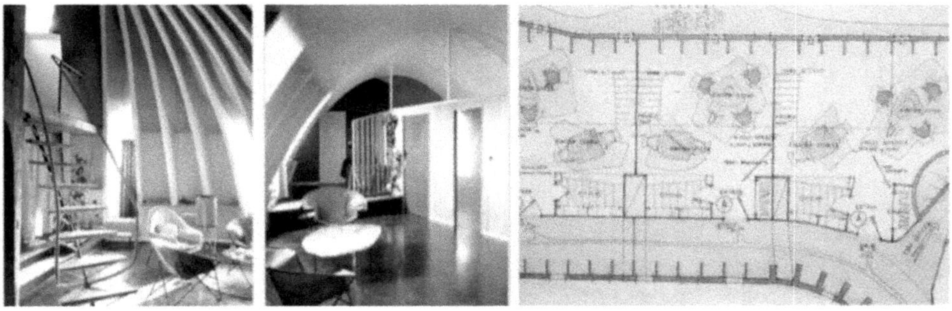

Interiors as Spatial Components

Some buildings and interiors owe their particularity to their spatial organization or in other words to the way spaces come together just like the Han typology developed by the Ottomans. Han buildings are characterized by the repetitive use of identical spaces called cells, which usually come together around a central space like a courtyard. Figure 4 shows one of the typical examples of this typology, Ali Pasa Han situated on the historic peninsula of Istanbul. The building offers different levels of interaction to the surrounding urban spaces with cells opening to streets and to the courtyard and becomes important with its functional competence thanks to its spatial organization. In this case interiors are not so impressive on their own but they create an overall spatial consequence and quality by the way they come together. While proposing reuse proposals for these kind of buildings, the role of interiors as spatial components need to be underlined by conserving their particular spatial organization. This necessitates preservation of the interior envelopes and the main concept of spatial circulation that connects them. In numerous recent reuse proposals, these repetitive cellular spaces are evaluated as restrictive because of their minimal dimensions that may not be efficient for activities and functions that require more space and consequently some interior walls are eliminated to maintain bigger spaces. However this kind of interventions break the concept of spatial repetition which is the most important factor of the building typology and ruin the cultural significance of the building. In the example of Ali Pasha Han the cell is the key element of architectural and spatial identity creating a unique repetitive pattern both with its form and function. (Saglar Onay, Minucciani, 2019).

Figure 4. Ali Pasa Han, Istanbul
(Author Saglar Onay's drawings).

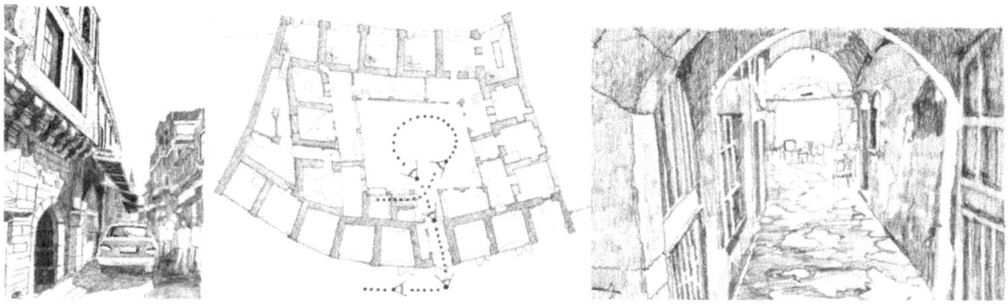

Interiors Perceptually Detached From the Architectural Envelope

A third case in Turin that, together with the nearby Palazzo del Lavoro, was the symbol of the celebrations of "Italia '61", is the Palazzo a Vela", a multi-purpose building originally called "Palazzo delle Mostre", designed by Franco Levi, Annibale and Giorgio Rigotti, owes its merit to a bold and harmonious, self-supporting structure, the sail to which it owes its name, which rests on the ground on only three points (with underground tie rods) expressing a lightness apparently incompatible with the reinforced concrete that constitutes it. The glazed walls of the perimeter, in addition to emphasizing the lightness of the structure, help to generate an interior space so large and bright as to appear as an exterior. (Saglar Onay, Minucciani, 2019).

Also in this case the extraordinary spatiality has been experimented in particularly suggestive exhibitions (like the one on Calder arranged by Renzo Piano in 1986) or to install functions that could hardly be hosted in an interior (an athletic track, a climbing gym). (Saglar Onay, Minucciani, 2019).

Nevertheless, also in this case the lack of strong functional projects made this structure difficult to perform until, after years of oblivion, a project of reuse was carried out in occasion of XX Winter Olympic Games in 2006. In 2003 Gae Aulenti Architect with Aurelio De Bernardi Engineer converted it as an ice rink. Unfortunately, the preservation of the shell was only partial: the reinforced concrete sail was saved while the boundless windows were removed. Without a perimeter curtain wall, the interior was completely erased and the sail became the simple "protective cover" of a second building with a radically different geometry and even strident with respect to its harmonious curve. A composition of red parallelepipeds emerges from the roof, humiliating the sail and depriving it not only of its function and autonomy, but also of its interior (Figure 5) (Saglar Onay, Minucciani, 2019).

Figure 5. Left and middle: "Palazzo a Vela" in its original version and right its reuse by Gae Aulenti Architect (2003)
(http://www.atlanteditorino.it)

Even if the structure has been physically preserved, it cannot be said that the space it created has been conserved. Interior, as detached from its architectural envelope, has not been considered in preservation issue: architecture is its space, and above all its interior. (Saglar Onay, Minucciani, 2019).

Interiors Reflecting Particular Aspects and Elaborate Detail

While changing the scale in observing interiors, details are the most volatile layer, extremely difficult to preserve in time. A further case study will show that details can be preserved as a significant part of architectural interior. Café Baratti di Milano in Torino is a historic company located in the heart of old Turin. Founded in 1858, it has remained in business all this time and was the supplier of the Royal House. (Saglar Onay, Minucciani, 2019).

When in 1872 the "Galleria dell'Industria Subalpina" (a pedestrian passage) was designed, it became an elegant and bright place for commerce and offices, and several commercial activities were moved there. In 1874 also Baratti & Milano decided to settle in the gallery becoming one of the busiest tea rooms in the city. Local newspapers celebrated this new shop, noting "sumptuousness and good taste", "mirrors of large surface area... artistic walnut woodcarvings... splendid gilding" (Dellapiana, Massabò Ricci, 2019).

In 1911, the fiftieth anniversary of national unification, the extension of the Baratti & Milano café was inaugurated: in fact, in 1909 the company had acquired further rooms and had entrusted the project to the architect Giulio

Casanova in collaboration with the sculptor Edoardo Rubino. They adopted the Art Nouveau language, conceiving the new rooms with great refinement and care in details also in the furniture and chandeliers.

In 1942 and 1943 the bombing did considerable damage demolishing some parts of the Subalpina Gallery: the Café was also partially damaged. The glass furniture was destroyed and the furniture in wood, marble, mirrors, chandeliers, fabrics and objects were ruined, but in 1944 the renovation of the cafeteria began and the accurate restoration testifies a positive approach towards this heritage. In 1985 the Ministry for Cultural and Environmental Heritage imposed protective restrictions on the spaces and on the furnishings, thus recognising their historical value. In 2003, the property carried out a thorough conservative restoration of the prestigious premises, bringing the furnishings, stuccoes, decorations, marbles, wooden inlays, the façade and the stained-glass windows back to their original splendour (Figure 6). (Saglar Onay, Minucciani, 2019).

Figure 6. Above: "Baratti & Milano" after war damages in 1944, and its previous appearance. Below: the interior as it appears today, and some details of the inlaid marble floor
(https://www.barattiemilano.it).

In this case, the primary role of interior architecture has been recognized in defining the historical and cultural value of a place: customers feel in this place the same atmosphere as a hundred years ago, thanks to the colors, materials and details - which characterize the interior architecture design that have been preserved with great care. Not only spatiality, therefore, is evaluated as a cultural value to be preserved, but also the refinements themselves become an integral part of cultural significance. (Saglar Onay, Minucciani, 2019).

Urban Interiors (Exterior Interiors)

Very similar, albeit with different results, is the example that concludes the case studies in Turin: a site linked to the great theme of the recovery/reuse of dismissed industrial buildings, which in effect today we call industrial Heritage. These are the warehouses of the former Teksid - Deltasider, industrial activities linked to the establishment of the FIAT car factory. Dismantled in the 1980s, they have been the subject of studies and national competitions for the redevelopment not only of the volumes but of the entire area, lapped by the river Dora that had been covered to increase exploitable surface. Some structures were too heavy, and the difficulty of eradicating the imposing pillars with deep foundations suggested leaving them as authentic ruins of industrial archaeology in a renovated urban park. While only the pillars of a shed were left behind, in a second one also the roof was saved, obtaining (as in the case of Palazzo a Vela) a protected space of considerable height and size. Below this, the most diverse events take place: from the "street food" event to football matches; from Muslin rituals to rock concerts (Figure 7). Unlike the Palazzo a Vela, the interior has still a spatial value, being transformed into a covered exterior, then protected, and perhaps it is more correct to say that it changed its identity and its scale: from private, monofunctional interior to urban, pubic and multifunctional exterior. (Saglar Onay, Minucciani, 2019).

The memory is saved, but at the same time it's been turned to the present. Interior spaces – very simple from the formal and functional point of view – preserve their identity, although as "ghosts" but still recognizable.

Figure 7. Different uses of new "urban interior": playground (www.lamiatorino.it); prayer on the occasion of the conclusion of Ramadan (https://torino.repubblica.it/cronaca/2012/08/19); "Kappa Futur Festival" Torino Summer Music Festival (www.gazzettatorino.it); street food" (www.mole24.it/2017/06/09).

A METHOD FOR EVALUATING SPATIAL POTENTIALS OF HERITAGE BUILDINGS AND INTERIORS

While analysing historic buildings of course it is very difficult to use the same method or framework for every building because all architectural structures have their own priorities and potentials to be underlined depending on their cultural significance. But it is important that each architect or designer can develop his own individual method in order to appreciate the quality embedded in historic buildings and recognize the underlying spatial issues, which are difficult to comprehend by an intuitive approach. Therefore analysing and evaluating historic buildings vary according to

- The cultural significance and heritage value of the building
- The priorities and approach of the architect
- The intention of the intervention

The original contribution of this part of the chapter is a framework for analysing existing buildings in order to determine the most compatible purpose and way of using them. The priority of the method is to understand the architectural and spatial potentials of the building and its components in order to convey these findings to the design process in adaptive reuse.

Adaptive re-use does not always concern the building as a whole. But even if adaptation proposes partial use, understanding the building as a whole is necessary. While the context of any architectural structure is the city or the landscape it is defined in, the context of every interior is the building it is situated in. So this work proposes a framework, which is composed of two main steps (Saglar Onay, 2017):

- Analysing the building as a whole
- Analysing interiors

Table 1. The framework for analysing heritage buildings and interiors

Analysing the building as a whole	**Analysing interiors**
Contextual Ties	Spatial Envelope
Tectonic Order and Material	Human Scale
Geometric Relations	Interior details, Furniture, Accessories
Spatial Organization	Spatial Atmosphere
Spatial Circulation	
Architectural Potentials	**Spatial Potentials**

(Saglar Onay, 2017).

Analysing the Building as a Whole

In order to understand a building, it is necessary to grasp the exterior forces that shape it. Buildings that are situated in urban context need to be analysed in terms of urban continuity. Therefore the relationship of building mass to the urban space in which it is defined in can be regarded as the initial phase of contextual analysis because generally urban space is one of the most important determinants of the architectural shell, which is an interface between the building and the city. Urban infrastructure also plays an important role in the determination of the buildings entrances and orientation. While analysing the relationship between the building and the urban space that surround it, it is necessary to look at the urban spatial envelope, predominant paths, density, important buildings, structures and landmarks around the building.

Another important component of analysing heritage buildings is material and tectonic order, which cannot be understood without the nature of materials used in these buildings. Material is much more than a follower of form. It is the strongest instrument of the architect in order to create a unique architectural language. Its structural capacity, physical and aesthetic aspects determine tectonic order and constructional principles. As tectonic order is accomplished by material, it is to be analyzed always in relation to material.

In most historic buildings especially in those of Renaissance the nature of material was the basic element that determined constructional principles and beauty was the result of the logical proportional relations and constructional principles that respected the nature of material. The most important deficiency of today's architecture and spatial design activities can be regarded as the detachment of construction from the realities of matter and craft. According to Meiss (1990), in architecture the question of truth and untruth refers to the relationship between form and construction. In this sense historic buildings expose the way they are structured in both elementary and holistic point of view.

Tectonic order arises from the relationship between architectural elements that come together with the principles of gravity. Gravity helps to define a hierarchy, which is vital for the consistency of the building.

Resolving geometries is another important step in evaluating historic buildings because all buildings are defined by the togetherness of basic or complex geometric forms and these geometries create spatial forces according to their articulation. The geometric relations of different parts of a building can be discussed in terms of building mass, façade, plan, section and interiors.

Spatial organization is defined by the relation ship of spaces that make up the building. In a building program spaces with different intentions and priorities come together. In this sense in most buildings there are spaces that are analogous or similar and there are spaces that differentiate from the rest. These differences define a hierarchy and every space is defined in the system by its role. Ching (1996), defines 5 organization types which are centralized organization, linear organization, radial organization, clustered organization and grid organization. Usually in historic buildings different spatial organization types are used together in order to define spatial hierarchy.

The last component of 'analysing buildings as a whole' is spatial circulation. Spatial circulation is the way or path that is predefined by the architect in order to create hierarchical access for all spatial components in a building.

While spatial organization is much more about the relationship of spaces, circulation is about the relationship of paths and spaces. In order to understand spatial circulation it is essential to consider the starting point, the layout and the architectural definition of the path.

Analysing Interiors

In analysing interiors of heritage buildings, the key element is the spatial envelope. The spatial envelope helps to define space by separating it from its surroundings. It is composed of surfaces with different roles: ground, walls and ceiling. The ground acts as the most important component of the spatial envelope because of its role as the supporter of human activity. According to Meiss (1990), the ground is the only indispensible component of space. On the other hand, walls, as the first surfaces to be perceived in space, have a different role with their power to orientate human activity. The openings on wall surfaces create forces from inside to outside and from outside to inside. The ceiling is the most challenging surface to be built and its most important role is to create spatial enclosedness. It also has the potential to orientate human activity depending on the form of the surface. The height of a room is much more important than the width and length because of its role in the determination of spatial identity.

The spatial envelope cannot be fully understood only by discussing geometric relations, openings and dimensions. Material is the physical reality of space and it is one of the most important determinants of tectonic order. The surfaces that define the spatial envelope of interiors also come together in tectonic hierarchy so material also determines this hierarchy by its strength and resistance. The color, transparency, opacity, texture and other superficial properties of material play an important role in the definition of spatial atmosphere.

The required hierarchy in historic interiors can only be achieved if the existing structural details, furniture, fittings, accessories, objects and their roles in the definition of spatial identity are discussed. Existing interior elements that contribute to the cultural significance of the building and interiors need to be preserved and utilized if possible. So, their physical properties and the way they interact with the spatial envelope needs to be analyzed. On the other hand, adaptation also necessitates the use of various new spatial elements in order to support human activity. So, interventions on historic interiors usually define new layers on existing spatial layers. These new layers are defined for shorter periods in order to adapt space to new uses and requirements. So, they need to be differentiated form existing layers in design language and material. These elements can be fixed or free standing according to the spatial properties and functional requirements.

While analyzing spatial potentials of interiors, the proportional relationship between human scale and the spatial envelope has vital importance as space has functional value only in relation to human body. The human being is always and necessarily conditioned in his life by his behavior in relation to a surrounding space (Bollnow, 2011). He interacts with space based on both his physical and sensorial capabilities. Space and spatial elements are understood only in relation to human scale and human activity.

The most intangible aspect of interior space is spatial atmosphere. When architecture is built, it is given physical form. But when human experience comes into scene, we start talking about spatial atmosphere. According to Unwin (2009), in our experience of architecture, basic architectural elements and the places they identify are modified by light, color, sound, temperature, air movement, smell and the qualities and textures of the materials used. Possible configurations of basic and modifying elements are probably infinite. Therefore as time goes on and spatial experience changes, the same space appears to be different. The spatial atmosphere changes continuously with the modifying elements of architecture. Therefore when we analyze interiors we need to take into account the effects of all these modifying elements in order to understand how they change. Sometimes these elements can be much more important than the physical definition of space as they stimulate all our senses.

Determining Architectural and Spatial Potentials

One of the most important objectives of spatial analysis in historic buildings is determining their architectural and spatial potentials in order to define a compatible use. At the end of analysis the architect needs to consider certain issues that will guide the intervention (Saglar Onay, 2017).

- the role of the building in its context / the role of interiors
- spatial restrictions, limitations
- spatial properties to be underlined
- factors that determine architectural and spatial identity
- values of cultural significance

Architectural or spatial role may refer to the use of a place and its contribution to a social system. Use means the functions of a place, including the activities and traditional and customary practices that may occur at the place (Burra Charter, 2013). Every building or structure can be evaluated as a part of a bigger spatial organization and it has an individual role in this system. This role can change in time or it can stay the same according to its location and spatial properties. Likewise, interiors also have roles in the spatial organization of a building. According to the spatial factors that have been discussed above, an interior might have an innate role that is hard to change.

In historic buildings determining spatial restrictions and limitations is very important in order to define the type and extent of the intervention. These restrictions may result primarily from tectonic logic and the fabric of the building. But interventions on historic buildings are not to be determined only by physical restrictions but also by a cautious approach that proposes spatial change only if necessary.

Especially in buildings with considerable historic and cultural background we talk about the togetherness of certain architectural and spatial values and layers. As discussed before analysis is looking at an entity by different filters and trying to understand the roles of different layers that make up the whole. In time as a result of damage, alteration or inappropriate interventions, a lack of hierarchy and correlation can be observed among these layers. In this case all layers that are of cultural significance need to be organised in hierarchy by underlining certain spatial properties that determine architectural and spatial identity.

The most important objective of analyzing historic buildings is to determine the values that contribute to the cultural significance of the building. According to the Burra Charter (2013), cultural significance may change over time with use as a result of new information. Every period, every different use has its signs and effects on the building itself. Therefore in analyzing historic buildings, layers from different periods needs to be evaluated with a contemporary point of view.

The notion of interior architecture in this concentric and holistic approach is to donate interiors with spatial elements that support human activity and enrich human experience. Historic buildings already have the potential to enrich human experience but they need to be adapted to contemporary needs and expectations in order to be integrated into contemporary life. They need to be a part of human experience. What is critical at this point is the analysis of all underlying issues and spatial values that are part of their cultural significance. Every intervention on historic buildings introduces new spatial elements and materials that can be regarded as new layers of adaptation. In this sense, in heritage buildings, the legibility of old and new layers increases the ability of communication of the building ensuring a meaningful spatial experience which will in turn enrich the quality of social activities and strengthen community well-being.

FUTURE RESEARCH DIRECTIONS

Future research at the intersection of well-being and cultural heritage needs to focus on the evaluation of existing heritage buildings used for social activities in order to understand the key aspects that contribute to community well-being. These evaluations can include survey studies as well as observations. At this point, other than architecture, planning and design contributions from psychology and sociology are necessary.

CONCLUSION

Interiors are the spaces we live in embracing most of our activities. So, if correctly evaluated, historic interiors can communicate all values of cultural significance making them part of spatial experience. In some examples interiors are considered less important but in fact authenticity necessitates conserving all layers that contribute to the building's identity. In the interiors we live in, while the architectural envelope although very important for urban totality is mostly something that we look at. Considering the continuity of spatial experience, interiors need to be considered as important as architectural envelope (Saglar Onay, Minucciani, 2019).

There is a difference between buildings that are conserved and those that simply remain. Conservation is also a question of spatial atmosphere. It's important to define appropriate uses and appropriate interventions that respect cultural significance: interior spaces (defined not only by their tangible envelope, but also by intangible features as light, temperature, smells…) maintain the "aura" of their original function, that's culturally as important as their physical shape. New functions that can be not consistent with it, create a sort of break in the memory (and actual conservation) line, while other, respecting the past, can emphasize the potential of shapes and spaces (Figure 8).

Figure 8. Preservation of the "atmosphere" associated with the original, religious function. Left: the interior of Selexyz Dominicanen Bookstore, 2007; right: mezzanine for the catering of Kruisherenhotel Maastricht, in the former Monastery of the friars of the Holy Cross, converted to a luxury hotel in 2011
Left: (https://mymodernmet.com/merkx-girod-selexyz-dominicanen-maastricht-bookstore-church); Right: (https://www.qlhotels.com/en/netherlands/limburg/maastricht/kruisherenhotel-maastricht).

As stated above, there is a strong connection between values of cultural significance and well-being. In this sense rather than individual well-being, it is important to consider community well-being. The existence of socially appreciated historic buildings and sites create a platform to come together and share common values of the society. This can be regarded as a booster of social well-being for all levels of the society. In this sense, the appropriate evaluation of historic buildings through adaptive reuse needs to be regarded as a very important contribution to well-being. Looking from this perspective well-being is not a concept that can be evaluated only for the individual but for also for the society. The understanding of community well-being is concretely supported by the historic environment and heritage buildings that reflect and respect common values of the society.

REFERENCES

APA dictionary of psychology. (n.d.). Washington, DC: American Psychological Association.

Ball, R. (1999). Developers, regeneration and sustainability issues in the reuse of vacant buildings. *Building Research and Information*, *27*(3), 140–148. doi:10.1080/096132199369480

Bollnow, O. F. (2011). *Human space*. Hyphen Press.

Brown, B., Perkins, D. D., & Brown, G. (2003). Place attachment in a revitalizing neighborhood: Individual and block levels of analysis. *Journal of Environmental Psychology*, *23*(3), 259–271. doi:10.1016/S0272-4944(02)00117-2

Bullen, P. A., & Love, P. E. D. (2010). The rhetoric of adaptive reuse or reality of demolition: Views from the field. *Cities (London, England)*, *27*(4), 215–224. doi:10.1016/j.cities.2009.12.005

Bullen, P. A., & Love, P. E. D. (2011). Adaptive reuse of heritage buildings. *Structural Survey*, *29*(5), 411–421. doi:10.1108/02630801111182439

Burden, E. (2004). *Illustrated Dictionary of Architectural Preservation*. McGraw Hill.

Carabi-Bescós, G. (2018). Signs of occupancy. 13 apartments in Gaudi's La Pedrera by Barba Corsini In *Proceedings of Fifth International Congress of Pioneers of Spanish modern Architecture: projects for inhabitance*, (pp. 52-71). Estonian Academy of Arts.

Ching, F. D. K. (1996). *Architecture, form, space and order*. John Wiley & Sons.

Chiorino, C. (2015). Cronistoria di una vita grama: il Palazzo del Lavoro dal 1961 a oggi. *Il Giornale dell'Architettura*. Retrieved from https://ilgiornaledellarchitettura.com/web/2015/09/20/cronistoria-di-una-vita-grama-il-palazzo-del-lavoro-dal-1961-a-oggi/

Deci, E. L., & Ryan, R. M. (1985). *Intrinsic motivation and self-determination in human behavior*. Plenum. doi:10.1007/978-1-4899-2271-7

Dellapiana, E., & Massabò Ricci, I. (2019). *Da bottega a società di capitali: Baratti & Milano in Torino*. Retrieved from https://www.barattiemilano.it/archivio-baratti

Department of Environment and Heritage (DEH). (2004). *Adaptive reuse*. Canberra: Commonwealth of Australia.

ICOMOS. (2013). *The Australia ICOMOS Charter for Places of Cultural Significance-The Burra Charter*. Retrieved from https://australia.icomos.org/

Kellert, R. (2005). *Building for life: Designing and understanding the human-nature connection*. Island Press.

Kyle, G., Graefe, A., Manning, R., & Bacon, J. (2004). Effects of place attachment on users' perceptions of social and environmental conditions in a natural setting. *Journal of Environmental Psychology*, *24*(2), 213–225. doi:10.1016/j.jenvp.2003.12.006

McCallum, D. (2007). Regeneration and the historic environment. In M. Forsyth (Ed.), *Understanding Historic Building Conservation* (pp. 35–45). Blackwell Publishing Ltd.

Meiss, P. V. (1990). *Elements of architecture*. Routledge.

Proshansky, H. M. (1978). The city and self-identity. *Environment and Behavior*, *10*(2), 147–169. doi:10.1177/0013916578102002

Rollero, C., & De Piccoli, N. (2010). Does place attachment affect social well-being? *European Review of Applied Psychology*, *60*(4), 233–238. doi:10.1016/j.erap.2010.05.001

Ruiz Millet, J. (1995). *Barba Corsini: Architecture 1953-1994*. Barcelona: Galeria H2O.

Ryan, R. M., & Deci, E. L. (2000). Self-determination theory and the facilitation of intrinsic motivation, social development, and well-being. *The American Psychologist*, *55*(1), 68–78. doi:10.1037/0003-066X.55.1.68 PMID:11392867

Saglam, N. (2015). *Interior layers and a case study in the Old Bedesten Cevahir* (Master Thesis). ITU Graduate School of Science Engineering and Technology.

Saglar Onay, N. (2014). Evaluating partial reuse of historical buildings as commercial interiors: A case study in Florence. *The International Journal of Architectonic, Spatial, and Environmental Design, 7*.

Saglar Onay, N. (2017). *Analysing spatial potentials of historic buildings*. Maggioli Editore.

Saglar Onay, N., & Minucciani, V. (2019). Interior space as communicator of cultural significance, In *Proceedings of XVII International Forum World Heritege and Legacy*. Roma: Gangemi Editore.

Sickels-Taves, L. B. (1999). *The use and the need for preservation standards in architectural conservation*. ASTM Publications. doi:10.1520/STP1355-EB

Stubbs, M. (2004). Heritage sustainability: Developing a methodology for the sustainable appraisal of the historic environment. *Planning Practice and Research, 19*(3), 285–305. doi:10.1080/0269745042000323229

Triandis, H. C. (1995). *Individualism and collectivism.* Westview.

Trucato, L. (n.d.). *La Pedrera, objeto de museo* (Bachelor thesis). Polytechnic of Turin.

Unwin, S. (2009). *Analyzing architecture.* Routledge. doi:10.4324/9780203880906

Wilkinson, S., & Reed, R. (2008). The business case for incorporating sustainability in office buildings: The adaptive reuse of existing buildings. In *Proceedings of 14th Annual Pacific Rim Real Estate Society Conference.* Kuala Lumpur: Pacific Rim Real Estate Society.

Wilkinson, S., Reed, R., & Kimberley, J. (2009). Using building adaptive reuse to deliver sustainability in Australia. *Structural Survey, 27*(1), 46–61. doi:10.1108/02630800910941683

Winter, J. (2007). Conserving buildings of the modern movement. In M. Forsyth (Ed.), *Understanding Historic Building Conservation.* Blackwell Publishing Ltd. doi:10.1002/9780470691823.ch8

Chapter 11
Conclusion:
New Perspectives for Well-Being

ABSTRACT

This last chapter discusses the need for interdisciplinary research and approaches for well-being. Therefore, it is very important to investigate the intersections of well-being with architecture, social sciences, medicine, as well as neuroscience. Recent well-being research shows that we have the amazing ability to reshape our brains, to change our affective mindsets, and to improve our well-being. On the other hand, it is necessary to understand cultural differences and its effects on well-being and to grasp the issue in accordance with the specific context. Therefore, the conclusion tries to understand how design and architecture can contribute to the way we build up well-being. Since interior space is one of the most important determinants of our everyday experiences, its role in well-being as a conscious construct needs to be the most important concern of spatial design.

INTRODUCTION

Interior environment needs to be evaluated as a very strong determinant of general well-being. This is also because interiors mark out a domain that is controllable in the first step by the designer and in the second step by the user. This control mechanism needs to be structured in such a way that interior design creates the appropriate stage for living and gives the necessary tools to the user to decide on his well-being himself.

DOI: 10.4018/978-1-7998-4231-6.ch011

Research into well-being and subjective experiences in developing countries is growing rapidly and represents a paradigm shift towards holistic, person-centered, and dynamic understandings of people's lives, which are nonetheless embedded in particular socio-cultural contexts (Boyden 2004; Gough et al. 2007; Camfield, Streuli et al. 2008). People's values, aspirations, and experiences of happiness or unhappiness are now measured directly within some large surveys of individuals and households rather than inferred from proxies such as income (Camfield, Crivello,Woodhead, 2009).

Well-being can be defined as the positive state of mind and body in relation to the complex system of interconnected components that built up living. By many psychologists well-being is seen as a result of the dominance of the positive effect on the negative effect. According to Deci and Ryan (2008), subjective well-being is most often interpreted to mean experiencing a high level of positive affect, a low level of negative affect, and a high degree of satisfaction with one's life. According to Bradburn (1969), as one of the first researchers in the field, an individual will be high in psychological well-being in the degree to which he has an excess of positive over negative affect and will be low in well-being in the degree to which negative affect predominates over positive. On the other hand, experimental studies confirmed that negative events activate our meaning system more effectively than positive ones (Baumeister, Bratslavsky, Finkenauer, Vohs, 2001) Negativity bias may be a response to a lack of felt social connectedness, which is an increasingly important problem in industrial societies. Generally, negativity bias refers to the fact that something negative has more impact on our thoughts, emotions and social interactions than something positive of equal intension (Hasler, 2016). And it is very important to note that negativity bias is one of the most consistent risk factors of mood and anxiety disorders. Regarding the evaluations of the interior environment about well-being requirements, the negative and positive conditions about the same spatial aspect is not similar. Users tend to pay more attention to aspects that have negative effect on their well-being. For example in the survey study run in Turkey and Italy (discussed in Chapter 5) the hierarchy of well-being requirements related to negative conditions was significantly different than the results of the question related to positive conditions. Although the answers for negative and positive conditions were arranged similarly (For example for ergonomic requirements the positive answer is 'having comfortable furniture/ equipment for activities' while the negative answer is 'having uncomfortable furniture/ equipment for activities), respondent's conception of hierarchy changed radically. As a result it was noted that the negative conditions about the same spatial aspect have more effect than the positive condition. People tend to evaluate negative situations in a more intensive manner. This finding arises as an issue that complicates the concept of well-being related to space. This might mean that eliminating the effect of negative distractions is more important than creating positive distractions. But lack of positive distraction might also lead to a lower stimulation of senses which might again result in low reported well-being. In this sense future research needs to focus on the hierarchy of negative and positive distractions as well as their relations with different dimensions of well-being related to requirements like social, ergonomic, sensory etc.

Well-being studies have been under the domain of social sciences for a long time. But today many other disciplines started to get involved in well-being studies, as it is clearly understood that social sciences alone cannot lead us to a comprehensive understanding of well-being. One of these disciplines is neuroscience known as the brain science, which studies the structure of the nervous system, and related cognitive processes, which are strongly connected to our spatial experiences. Also medicine has broadened the concept of health in an understanding that embraces well-being because it is understood that health and well-being are strongly related concepts getting into each other. Similarly studies in social sciences have started to broaden the understanding of the concept trying to create platforms for interdisciplinary research. So, we can say that well-being studies require an interdisciplinary approach in order to go further.

THE INTERSECTIONS WITH OTHER DISCIPLINES

Neuroscience

Studies in the intersection between neuroscience and space mostly use virtual environments because they let users navigate and interact with computer-generated three-dimensional environments in real time, allowing for the control of complex stimuli presentation. On the other side, in the interaction between virtual space and human there are many questions to be answered. These questions are mostly about the structure of human cognitive processes, motion sickness concerns, and computing parameters (Tarr, Warren, 2002).

Our understanding of the neural basis of spatial perception and memory has been advanced by recordings of neural activity within the hippocampus as rats explore an environment. Such studies reveal that individual neurons respond when the animals move to certain locations in that environment, with different patterns of neural activity corresponding to different locations. Unlike neurons in sensory areas of the brain, these "place" neurons are not activated by any one type of stimulus, such as a visual feature, or a sound, or a smell, but rather by the combination of features that serve to define the animal's internal sense of place (Nakazawa et al., 2004; Sternberg & Wilson, 2006). Sternberg and Wilson discuss that place neurons are dependent upon a strong sense of orientation within the environment, and different environments are associated with different patterns of neural activity. Because our sense of "place" may be defined by the activity of these neurons, understanding the factors that influence them may help to elucidate how we are impacted by architectural design elements. Another research indicates that the way in which memories of places are formed during exploration of an environment and then later retrieved show that memory of place is not composed of isolated locations but rather sequences of connected locations (Lee and Wilson, 2002).

An abundance of neuroscientific evidence indicates that the subjective experience of happiness, pleasure, rewards, and general well-being depends on shared or overlapping networks of interacting subcortical and cortical regions of the brain. Given that the brain circuits supporting positive emotions and rewards are remarkably flexible, we have the amazing ability to reshape our brains, to change our affective mindsets, and to improve our well-being (Dolcos, Moore, Katsumi, 2018). In this sense if we can understand the ways brain processes well-being, it might be enhanced through training and the space for well-being can be interpreted in a way that helps this training of the brain.

Another subfield of neuroscience that can be regarded as important for architecture and design is Neuroaesthetics. Neuroaesthetics can properly be viewed as a subfield of cognitive neuroscience, given that it studies a form of human cognition and behavior using a combination of techniques and methods from neuroscience and cognitive science, bringing together the cognitive and neural levels of explanation (Churchland & Sejnowski, 1988; Gazzaniga, 1984; Pearce, Zaidel, Vartanian, Skov, Leder, Chatterjee, Nadal, 2016). Most uniquely, humans have many prominent higher order, abstract or cultural pleasures, including personal achievement as well as intellectual, artistic, musical, altruistic, and transcendent pleasures. While the neuroscience of higher pleasures is in relative infancy, even here there seems overlap in brain circuits with more basic hedonic pleasures (Frijda 2010; Harris et al. 2009; Leknes and Tracey 2010; Salimpoor et al. 2011; Skov 2010; Vuust and Kringelbach 2010; Berridge, Kringelbach, 2011). Studies indicate that aesthetic qualities of architecture have an impact on our mood, cognitive functioning, behavior, and even mental health (Adams, 2014; Cooper, Burton, & Cooper, 2014;Hartig, 2008; Joye, 2007). This evidence coincides with a flourish of interest in the intersection of neuroscience and architecture (Dance, 2017; Robinson & Pallasmaa,2015; Mallgrave, 2010; Eberhard, 2008). However, relatively little work has been conducted on the neuroscience of architecture (Coburn, Vartanian, Chatterjee, 2017).

According to Hasler, the important influence of individual and cultural influences on positive health, is a huge challenge for the measurement of well-being. In addition, psychometric instruments that are clinically useful may not be appropriate to study the neurobiology of well-being. On the other hand, there are scientists that concentrate on the potentials of Asian wisdom in combination with modern neuroscience. Buddhism-inspired research in meditation and mindful training is reported to increase neuronal activity exactly in these two regions, i.e. the anterior cingulate cortex and the anterior insula. This means that mindfulness-based psychotherapies work across diagnostic categories and show high efficacy to reduce the risk of relapse (Hasler, 2016).

Another important emerging research in the intersection of neuroscience and well-being is the increasing use of social media. In the age of social media, people seem to be more connected to each other following what the others are doing, seeing their pictures etc. But on the other hand as time spent on social media eats more and more into the time we used to spend face to face with real people, it will be important to know what changes, if any, social media induces in users' functioning. Although social media can elicit positive consequences, exploring the effects of social media on the brain is especially relevant considering that social media use can develop into a behavioral addiction, and can contribute to poor academic performance, job loss, and declines in well-being (Verduyn, et al., 2015; Barchard & Williams, 2008; Meshi, Tamir, Heereken, 2015).

Social Sciences

From different branches of social sciences, there are emerging views about well-being which criticizes the individuality of the concept. According to Hasler (2016), well-being should be more than individual well-being. Similarly Oliver (2020), discusses that the more we see ourselves as discrete entities, the more likely we are to feel isolated and lonely and to show "selfish" behaviours. This leads to mental disorders and negatively effects well-being.

There are also more research studies that try to combine different traditions of well-being. British sociologist Paul Dolan discuss that human beings usually do not want to just feel good, but also want to feel that we are doing good. So, it is necessary to balance and combine these two needs to enhance overall well-being. This view also supports that the hedonic and eudaimonic traditions are both effective on general well-being.

We know that architecture can effect the way people behave and interact with each other. The social and psychological dimension of well-being can be enriched by architectural solutions. In the previous chapters we have discussed that even in hospital environments individual rooms and being isolated from the others, have negative effects on both mental and physical health. The spatial aspects of social well-being are related to qualities that enrich social life. In this sense, buildings should provide the chance for social encounters at desired level for everyone. Moreover they should support and encourage social activities creating a balance between public, semi-public and private places. The quantity and quality of social connections (e.g. talking and listening to family or strangers) correlate with reported well-being as well as physical health.

Anthropology and Ethnography

The application of the framework to different situations shows that well-being requirements evolve during people's lifespan but above all they depend on the cultural context. They depend on the relationship that people have with environmental conditions, especially natural and climatic ones. Therefore, the habits of life, strongly dependent on culture, imply customs, rituals, social practices that strongly affect spaces, especially interiors. Flexible rather than large spaces are then required; interconnected rather than intimate; interiors strongly connoted from a decorative rather than more neutral point of view. Anthropology cannot afford to ignore well-being, because there is no single pursuit of happiness, but multiple: "it means different things in different places, different societies, and different cultural contexts". Well-being is thus differently conceived, expressed and experienced (Matheus and Izquierdo, 2009).

Moreover, cultures are dynamic, so the concept of well-being itself evolves within them. An ethno-anthropological approach to well-being would be of great value by coming together with the discipline of architecture. To understand the cultural meanings of well-being is to understand a society in its different aspects, include social and political ones, that are reflected in practices and interpretations. (Adelson, 2009)

According to Colby, who wonders whether the measure of cultural well-being is possible or even desirable, the reason why anthropologists have often avoided going into the subject of well-being is precisely in the cultural relativism, rendering comparison impossible. Other topics are closely related to it: the locality of the culture, intended as geographical, physiological or even metaphysical (Colby, 2009).

"Culture", notwithstanding, remains a vague concept. At the same time, the double and reciprocal relationship that links cultural belonging to the natural and built environment must be the subject of much more systematic and in-depth studies, with particular reference to the qualities of interior spaces.

Medicine

Health is a dynamic state of well-being emergent from conducive interactions between an individual's potentials, life's demands, and social and environmental determinants. It results throughout the life course when an individual's potentials and social and environmental determinants suffice to respond satisfactorily to the demands of life. Life's demands can be physiological, psychosocial, or environmental, and vary across individuals and contexts, but in every case unsatisfactory responses lead to disease (Bircher, & Hahn, 2016). So, the environment is a very important determinant of health. It also plays a role in training our bodies coping with problems. From the perspective of medicine a person has two potentials, the first is given or innate and the second is achieved in time through the interaction with the physical and social environment. According to Bircher and Hahn, interactions in families, schools, and communities play a pivotal role for children and adolescents in supporting their acquisition of knowledge and skills and personal development. Even in adulthood the development of the potential continues. The social determinants of health have a very important enabling or inhibiting function. Integration and participation in the society is of outstanding importance for each individual. The society also provides many other helpful possibilities, such as work, food shops, health information, social security, and a health care system (Bircher, & Hahn, 2016). So, environmental determinants of health have impacts on the two human potentials and on the demands of life. These determinants of health begin from the biosphere and proceeds in concentric levels until interior space and human body. On the other hand the environmental dimension is not detached from social dimensions of health. Attitude, culture, willingness to learn theoretically and from experience, and the possibility to grow with new challenges will have major influences on health. In principle every person has considerable influence on his own potential. If so, everyone might be considered as an entrepreneur of his or her own health. This idea as expressed in the Meikirch model is now being practiced successfully in indigenous villages in Odisha, India (Samal, 2014; Bircher, & Hahn, 2016).

Departing from the assumption that everyone is an entrepreneur of his own health, we can also assume that everyone can effect his own well-being and build well-being for themselves. The spatial and architectural dimension of well-being is very important at this step. As discussed in the previous chapter in both hedonic and eudaimonic tradition, space can significantly contribute to well-being.

FUTURE RESEARCH DIRECTIONS

Future research at the intersection of well-being and other disciplines needs to focus on the evaluation of existing living environments in order to understand the key aspects that contribute to individual and community well-being. These evaluations can include survey studies as well as observations. At this point, other than architecture, planning and design contributions from medicine, psychology, sociology, anthropology, ethnography and neuroscience are necessary.

CONCLUSION

This last chapter of the book aimed to underline the main concepts discussed through the book trying to look at intersections between architecture and other disciplines that can make specific contribution the concept of well-being related to space and architecture. This interdisciplinary way of thinking and research is extremely important for well-being because at the moment it is not possible to understand the nature of well-being with all its dimensions. Although every discipline has its own ways of explaining the value, meaning, components and promoters of well-being, it is obvious that there is much more to discover.

The general findings of the book "Well-Being Design and Frameworks for Interior Space" are listed below:

Well-being research in general is dominated by two different traditions. The hedonic tradition gives importance to constructs such as happiness, positive affect, low negative affect, and satisfaction with life while the eudaimonic tradition concentrates on the positive psychological functioning and human development. Regarding the importance of design from the perspective of both traditions, design can create the right tools with which a person can develop himself and find the right ways to perform his actions for his specific contribution and design can also promote pleasurable experiences and block undesired experiences. Moving to the role of interior design we can say that interiors can provide the right environment in which people can be free to perform the actions they chose in the way they want and they can appreciate pleasurable experiences as an outcome of their interaction with space. In this sense the interior environment needs to be evaluated as a very strong determinant of general well-being.

The concept of good life has not been similar through history. Exploring architectural clues at different times we can see that the concept of the good life and happiness changed, evolved sometimes as a response to previous conceptions, sometimes as a response to societal and technological changes and sometimes as an evaluation of the offerings of nature. In time the role of the individual and its independence gained more importance while in ancient times the individual's happiness was regarded as a phenomenon more related to the society.

In this book also the certification systems about well-being in architecture have been examined. Well-being is a holistic concept, made of complex interactions among several parameters. Actually, involving also psychological needs and individual lifestyles, well-being is differently perceived in different contexts. The typical approach of Interior Architecture entails a strict relation with clients leading to a real customization, then it should be more considered in developing international Certifications: they are instead targeted to a global audience, tending to be deterministic while in this field the subjective component is strong. In the system of certifications, in fact, the quality of interior spaces intended in their architectural and spatial connotations is not considered, then these systems risk mortifying architecture in its truest essence. Interior architecture is too often considered as a practical discipline, instead of a technical one and (before being technical) it must be cultural and theoretical: a discipline of research, practice and education.

While we consider wellbeing in interiors, the most important contribution can be about structuring an understanding of design that methodologically supports the concept. After understanding user's needs and defining the requirements, the designer needs to get involved in the design process which is generally a multi-criteria design task because of the complexity and subjectivity of wellbeing needs. In this sense, it becomes extremely important to determine the design approach that will guide the designer to the design output. There are already different design approaches that have been developed to contribute to the quality of life of users and inhabitants. A really holistic approach to interior design is very challenging to achieve, and research can only be experimental. But every design approach discussed in the book handles the issue of design from different perspectives and has the potential to contribute to the understanding of well-being related to space. They all have different focuses to improve the spatial quality of our living environments. Many of them have been developed as a response to problematic environmental situations. Inclusive design, in this sense, can be seen as a response to environmental injustice. It tries to create a common ground for all user groups that share the environment and its contribution becomes more significant considering especially public spaces. Likely supportive design aims to eliminate the negative effect of having health problems and tries to give relief especially by developing the psychological state of users. Approaches based on the positive effect of nature like biophilic and biomimetic design can also be regarded as a response to the detachment of our living environments from nature, trying to take advantage of natural stimulators and processes. Human centered design focuses on testing through prototypes and generative design takes the advantage of computer-designer interaction in order to create multiple design solutions.

The well-being framework is an important contribution of this book as it tries to structure all the requirements that contribute to well-being related to space. The framework proposal, consisting of contextual, functional, psychological, social, ergonomic, sensory requirements and aesthetic requirements as basic design criteria, aims to support both theoretical and practical activities regarding well-being in all living environments, as a crucial component of sustainability. Connection to context has an important role as the factor that makes every interior unique with its location and relationship to the rest of the world. Functional requirements are concerned with creating spatial solutions that support human activities. Psychological requirements are more indirect as they deal with the outcomes of the interaction between human and space. Social requirements deal with the spatial opportunities for the level of interaction between people. Sensory requirements are concerned with both quantitative factors like heating, illumination, acoustics, sound control etc. and their reflections on human senses. Aesthetic requirements are also important as they refer to the integrity and harmony of all spatial elements that can be pleasing to human senses and can promote well-being. Ergonomic requirements are more concrete and they refer to the relationship between all interior components and human body and capabilities.

The examination of well-being through the framework also gives the possibility to analyse the effects of different requirements and their hierarchic relations in different groups and different cultures. The application of the framework to different living environments helps to understand the different nature of well-being related to spaces with different purposes. In this sense survey studies in home environments and school environments has shown the importance and hierarchy of well-being requirements giving important clues for future design approaches. The results also show that well-being requirements evolve during people's lifespan but the most significant result concerns the differences between cultures. Although all of the well-being requirements are important to fulfil the necessary level of well-being, the ways of achieving these goals differ according to different contexts and understandings of users. This proves the generally accepted statement that designers and architects cannot handle design as a global issue and in every specific context there exists some general inclinations and some various tendencies of different age groups. Therefore, further research needs to focus on the relation of well-being and age in different cultural contexts to study the significant role of culture on well-being requirements.

Another important issue discussed in the book is the contribution of vernacular architecture to well-being. Through the examination of vernacular homes, we have seen that the requirements defined in the well-being framework all find their spatial correspondences. Today we are very far away from this point, as we designers and architects tend to create architectural solutions without a proper examination of ways of living. It is also a reality that the difference between cultures and lifestyles is not as significant as before but they are adapted to the advancements in technology and science. In this sense the mission of design and architecture is to understand this adaptation process so that the contemporary ways of living and thinking can find their correspondences in new living environments. Future research needs to discuss how new architectural solutions can respond to well-being needs in a natural way as we can see in vernacular examples and how we can develop an updated and contemporary understanding of vernacular architecture. This is also because our present ways of living is excessively impactful on the environment and living environments are no more extensions of nature. Conservation of vernacular architecture is an important issue to inspire future generations teaching us how to be integrated with the environment by taking the advantage of local sources. This understanding has to be a part of architecture and design education in order to discuss their exceptional design approaches and to adapt them to new living environments. Therefore conservation of vernacular architecture as a lesson to preserve a sort of "biodiversity" in lifestyles can contribute greatly to design education, research and practice. Vernacular houses represent an irreplaceable value for the interior architecture body of knowledge as they are greatly connected to cultural aspects, different from society to society, far away from the impacts of "Global" or "International Style" that create similar solutions in different contexts.

Regarding the potential of adaptive reuse, the existence of historic buildings and sites create a platform to come together and share common values of the society. This can be regarded as a booster of social well-being for all levels of the society starting from the people who can share memories about the past of these buildings and also including the younger generations who can explore concrete values about the society they live in. The existence and sharing of these important cultural values strengthen social ties and belonging. In this sense, the appropriate evaluation of historic buildings through adaptive reuse needs to be regarded as a very important contribution to well-being. Looking from this perspective well-being is not a concept that can be evaluated only for the individual but for also for the society. Individual well-being is not a concept independent from the community. The understanding of community well-being is concretely supported by the environment that respects common values of the society. In this sense adaptive reuse of buildings for social purposes is strongly linked with community well-being, which in turn directly affects individual well-being.

Architecture for well-being is a difficult challenge to achieve. Both well-being and architecture are complex structures to define and understand. Today we know that well-being is an inner state of human, which can only be reported but not concretely measured. But it is the same for architecture. Even it is a concrete aspect to be experienced, the perception and cognition of space and architecture is related to human nature. Also following the very complex openings that architecture has in this period with regard to humanities, it has to be particularly underlined that space is always "emotional", and it is never "neutral".

REFERENCES

Adams, M. (2014). Quality of urban spaces and well-being. In *Well-being and the environment* (Vol. 2, pp. 249–270). John Wiley & Sons.

Anderson, N. (2009). The shifting landscape of cree well-being. In G. Matheus & C. Izquierdo (Eds.), *Pursuit of happiness. Well-being in anthropological perspective* (pp. 109–124). Berghahn Books.

Barchard, K. A., & Williams, J. (2008). Practical advice for con-ducting ethical online experiments and questionnaires for United States psychologists. *Behavior Research Methods, 40*(4), 1111–1128. doi:10.3758/BRM.40.4.1111 PMID:19001403

Baumeister, R. F., Bratslavsky, E., Finkenauer, C., & Vohs, K. D. (2001). Bad is stronger than good. *Review of General Psychology, 5*(4), 323–370. doi:10.1037/1089-2680.5.4.323

Berridge, K.C., Kringelbach, M.L. (2011). Building a neuroscience of pleasure and well-being. *Psych Well-Being, 1*(3).

Bircher, J. (2005). Towards a dynamic definition of health and disease. *Medicine, Health Care, and Philosophy, 8*(3), 335–341. doi:10.100711019-005-0538-y PMID:16283496

Bircher, J., & Hahn, E. G. (2016). Understanding the nature of health: New perspectives for medicine and public health. Improved well-being at lower costs. *F1000 Research, 5*, 167. doi:10.12688/f1000research.7849.1 PMID:27134730

Bircher, J., & Kuruvilla, S. (2014). Defining health by addressing individual, social, and environmental determinants: New opportunities for health care and public health. *Journal of Public Health Policy, 35*(3), 363–386. doi:10.1057/jphp.2014.19 PMID:24943659

Boyden, J., & de Berry, J. (Eds.). (2004). *Children and youth on the front line: Ethnography, armed conflictand displacement*. Berghahn Books.

Bradburn, N. (1969). *The structure of psychological well-being*. Aldine.

Camfield, L., Crivello, G., & Woodhead, M. (2009). Well-being research in developing countries: Reviewing the role of qualitative methods. *Social Indicators Research, 90*(1), 5–31. doi:10.100711205-008-9310-z

Camfield, L., Streuli, N. & Woodhead, M. (2008). *Children's well-being in contexts of poverty: Approaches to research, monitoring and participation.* Young Lives technical note no. 12.

Coburn, A., Vartanian, O., & Chatterjee, A. (2017). Buildings, beauty and the brain: A neuroscience of architectural experience. *Journal of Cognitive Neuroscience, 29*(9), 1521–1531. doi:10.1162/jocn_a_01146 PMID:28493809

Colby, B. N. (2009). Is a measure of cultural well-being possible or desirable? In G. Matheus & C. Izquierdo (Eds.), *Pursuit of happiness: Well-being in anthropological perspective* (pp. 45–64). Berghahn Books.

Dance, A. (2017). Science and culture: The brain within buildings. *Proceedings of the National Academy of Sciences of the United States of America, 114*(5), 785–787. doi:10.1073/pnas.1620658114 PMID:28143973

Deci, E. L., & Ryan, R. M. (2008). Hedonia, eudaimonia, and well-being: An introduction. *Journal of Happiness Studies, 9*(1), 1–11. doi:10.100710902-006-9018-1

Dolcos, S., Moore, M., & Katsumi, Y. (2018). Neuroscience and well-being. In E. Diener, S. Oishi, & L. Tay (Eds.), *Handbook of well-being*. DEF Publishers.

Eberhard, J. P. (2008). *Brain landscape: The coexistence of neuroscience and architecture*. Oxford University Press.

Gough, I., McGregor, J. A., & Camfield, L. (2007). Introduction: Conceiving well-being in development contexts. In I. Gough & J. A. McGregor (Eds.), *Well-being in developing countries: New approaches and research strategies*. Cambridge University Press.

Hartig, T. (2008). Green space, psychological restoration, and health inequality. *Lancet, 372*(9650), 1614–1615. doi:10.1016/S0140-6736(08)61669-4 PMID:18994650

Hasler, G. (2016). Well-being: An important concept for psychotherapy and psychiatric neuroscience. *Psychotherapy and Psychosomatics, 85*(5), 255–261. doi:10.1159/000447268 PMID:27509027

Joye, Y. (2007). Architectural lessons from environmental psychology: The case of biophilic architecture. *Review of General Psychology, 11*(4), 305–328. doi:10.1037/1089-2680.11.4.305

Lee, A. K., & Wilson, M. A. (2002). Memory of sequential experience in the hippocampus during slow wave sleep. *Neuron*, *36*(6), 1183–1194. doi:10.1016/S0896-6273(02)01096-6 PMID:12495631

Mallgrave, H. F. (2010). *The architect's brain: Neuroscience, creativity, and architecture*. Wiley-Blackwell. doi:10.1002/9781444317275

Matheus, G., & Izquierdo, C. (2009). Introduction anthropology, happiness, and well-being. In G. Matheus & C. Izquierdo (Eds.), *Pursuit of happiness. Well-being in anthropological perspective* (pp. 1–20). Berghahn Books.

Meshi, D., Tamir, D. I., & Heekeren, H. R. (2015). The emerging neuroscience of social media. *Trends in Cognitive Sciences*, *19*(12), 771–782. doi:10.1016/j.tics.2015.09.004 PMID:26578288

Nakazawa, K., McHugh, T. J., Wilson, M. A., & Tonegawa, S. (2004). NMDA receptors, place cells and hippocampal spatial memory. *Nature Reviews. Neuroscience*, *5*(5), 361–372. doi:10.1038/nrn1385 PMID:15100719

Pearce, M. T., Zaidel, D. W., Vartanian, O., Skov, M., Leder, H., Chatterjee, A., & Nadal, M. (2016). Neuroaesthetics: The cognitive neuroscience of aesthetic experience. *Perspectives on Psychological Science*, *11*(2), 265–279. doi:10.1177/1745691615621274 PMID:26993278

Robinson, S., & Pallasmaa, J. (Eds.). (2015). *Mind in architecture: Neuroscience, embodiment, and the future of design*. The MIT Press. doi:10.7551/mitpress/10318.001.0001

Samal, S. (2014). Implementation of the Meikirch Model in Odisha, India. *Journal of Public Health Policy*, *35*(3), 387–391. doi:10.1057/jphp.2014.22 PMID:24943668

Sternberg, E. M., & Wilson, M. A. (2006). Neuroscience and Architecture: Seeking Common Ground. *Cell*, *12*(2), 239–242. doi:10.1016/j.cell.2006.10.012 PMID:17055420

Tarr, M. J., & Warren, W. H. (2002). Virtual reality in behavioral neuroscience and beyond. *Nature Neuroscience*, *5*(S11), 1089–1092. doi:10.1038/nn948 PMID:12403993

Verduyn, P., Lee, D. S., Park, J., Shablack, H., Orvell, A., Bayer, J., Ybarra, O., Jonides, J., & Kross, E. (2015). Passive Facebook usage undermines affective well-being: Experimental and longitudinal evidence. *Journal of Experimental Psychology. General*, *144*(2), 480–488. doi:10.1037/xge0000057 PMID:25706656

About the Authors

Valeria Minucciani is Architect and Associate Professor in Interior Architecture at Polytechnic of Turin, where she is Coordinator of Master in Interior, Exhibit & Retail Design and is member of the Teaching Board of School of Specialization in Architectural and Environmental Heritage. She is Coordinator in Piedmont for National Association of Small Museums (APM). She deals with Interior space, Museography, and Cultural Heritage enhancement and communication, according to different approaches. She studied in particular archaeological and religious museums, the ICTs role in enhancement of Cultural Heritage and the accessibility to culture. On these issues she's Author of several books, chapters and papers in national and international context. Her interests are currently focusing on disciplinary contamination between Interior spaces and neuroscience.

Nilüfer Saglar Onay is an architect and academician in the field of architecture and interior design. She got her B.Sc. (1999) in Architecture, M.Sc. (2001) and PhD (2010) in Architectural Design. She has been a visiting scholar at SUPSI, Switzerland in 2008, at UNIFI, Florence in 2010- 2011, at POLIMI, Milan and POLITO, Turin from 2015 until present. She started her professional academic career in Istanbul Technical University in 2004 as a research assistant and got the title of associate professor in 2015. She has contributed to several architectural projects and restoration projects in Istanbul. She has numerous national and international publications including journal articles, books, book chapters and conference papers. Her main research areas are public interior spaces, spatial evaluation models, residential architecture, adaptive re-use and well-being in interior design.

Index

H

I

L

M

N

P

Q

R

S

T

U

V

W

www.ingramcontent.com/pod-product-compliance
Ingram Content Group UK Ltd.
Pitfield, Milton Keynes, MK11 3LW, UK
UKHW050946290926
15470UKWH00005B/1676

9 781799 842316